Digital
DOMINANCE

Digital
DOMINANCE

Winning in a Socially Networked World

JAMES JAY CARAFANO

TEXAS A&M UNIVERSITY PRESS

COLLEGE STATION

∞ This paper meets the requirements of ANSI/NISO Z39.48-1992 (Permanence of Paper).
Binding materials have been chosen for durability.
Manufactured in the United States of America.

Library of Congress Cataloging-in-Publication Data

Names: Carafano, James Jay, 1955–author.
Title: Digital dominance: winning in a socially networked world / James
    Jay Carafano.
Description: First edition. | College Station: Texas A&M University Press,
    [2026] | Prologue: What Hath Wiki Wrought?—Backstory—The Story So
    Far—How to Win—How to Build Networks—Drawing a Crowd—An
    Offer You Can't Refuse—Becoming Forrest Gump—Failure to
    Communicate—How to Beat Networks—You're Connected—Keep the
    Towels—Bad for Breakfast—Crush Your Enemies—World of Networks
    —To Win Such a War—The World's a Stage—Guns and Butter—
    Emergent—Epilogue: Thinking the Future. | Includes bibliographical
    references and index.
Identifiers: LCCN 2025023288 | ISBN 9781648432897 (trade paperback) | ISBN
    9781648432903 (ebook)
Subjects: LCSH: Digital divide. | Information society—Political aspects. |
    Online social networks—Political aspects. | Metaverse—Political
    aspects. | Internet governance—Political aspects. | Internet and
    terrorism. | National security—Computer networks. | Cyber intelligence
    (Computer security) | Computer networks—Access control.
Classification: LCC HM851 .C2745 2026 | DDC 303.48/33–dc23/eng/20250615
LC record available at https://lccn.loc.gov/2025023288

Published by Texas A&M University Press
John H. Lindsey Building, Lewis Street
College Station, TX 77843
www.tamupress.com

This book complies with the EU General Product Safety Regulation (GPSR) (Regulation (EU) 2023/988). For regulatory inquiries and product safety matters within the EU, contact our authorized representative:
Mare Nostrum Group B.V.
Mauritskade 21D
1091 GC Amsterdam
The Netherlands
Email: gpsr@mare-nostrum.co.uk
Website: https://mngbookshop.co.uk/

This publication has undergone a risk assessment and meets applicable safety standards. Traceability identifiers: ISBN, batch number, and publisher contact details are provided for compliance. No hazardous materials or components are included in this product.

CONTENTS

Digital
DOMINANCE

PROLOGUE
WHAT HATH WIKI WROUGHT?

Zabihullah Mujahid had something to say: "There will be no discrimination against women" in the newly established Islamic Emirates of Afghanistan.[1]

When the Taliban returned to power following the chaotic withdrawal of American forces in August 2021, the world saw Zabihullah Mujahid's face for the first time. For years, he had been the group's shadowy public voice. Not one authenticated photo existed. Afghanistan intelligence doubted he was a real Afghan—or even a real person at all.[2]

But no ghost addressed the international press. Here was the public face of the "new" Taliban. The heavily bearded spokesperson garbed in traditional Pashtun dress smiled. He offered assurances on the new regime's reasonableness. Before American forces had driven them out of the country twenty years before the Taliban had ruled with a brutal hand, enforcing an extreme, harsh version of sharia that included the subordination of women and severely restricted personal freedoms. To make a point that the times had changed, he took his first question from a female journalist.

Not only were Zabihullah Mujahid's remarks from the August 19 press conference widely reported, but they were also echoed on forums such as Twitter (now X, which announced it would only block content that violated the company's rules and policies), where Zabihullah Mujahid already boasted more than three hundred thousand followers. In the following weeks, the Taliban followed up with a flood of tweets, pictures and videos, and burka-shrouded supporters (who all may or may not have even been women) meeting and marching in support of the fundamentalist order. At least one thing was clearly different: the Taliban had become wicked better at engaging international broadcast media and establishing a presence on digital social platforms.[3]

This effort had little to do with engaging the Afghan people. In 2021, globally about 60 percent of the population had access to the internet. According to the World Bank, in Afghanistan only about 13 percent of the population could get online. Afghanistan is a rare desert in the modern internet world. Zabihullah Mujahid's narrative was geared to impress the rest of the planet, where the internet spread like wildflowers.

As effective social networking online, the effort was a joke. Zabihullah Mujahid was unlikely to convince anyone that the "new" Taliban would prove any less brutal than the "old" Taliban, as facts would later prove. The "new" Taliban, for instance, have the worst record in the world for treating women with humanity and dignity. Two years after the Taliban invasion, Georgetown University's Index of Women, Peace, and Security (2023 edition) ranked Afghanistan 177 out of the 177 countries rated, with a score barely above zero. Women are banned from education and the workplace as well being employed in nongovernmental and international organizations. The Taliban's failure to convince us they were going to be kind and cuddly, however, had nothing to do with the reality of who they are or not knowing how to manipulate technology. During the war years, their leadership proved to be anything but a bunch of techno-Neanderthals. In battle, they used encrypted messaging in the Facebook-owned WhatsApp to coordinate attacks. But that is not the same as using social networking as an effective weapon.

The Taliban may have conquered the country, but they quickly lost the social networking war. From almost day one, Taliban disinformation on human rights was canceled out by a steady stream of grainy videos: women and journalists being beaten and abused and former government officials hunted down and tossed into the trunk of a car.

While Zabihullah Mujahid showed he could deliver far slicker propaganda than the old regime, he flunked at following the rules for being a social networking ninja. If the Taliban had to be dependent on tools like YouTube instead of gun tubes, the leadership would still be vacationing in Quetta (Pakistan) instead of ruling in Kabul (Afghanistan).

No insult to Zabihullah Mujahid. What little biography we have of him suggests he is a serious person. From what we know, he is an ethnic Pashtun from eastern Afghanistan, middle aged, and well educated. After serving as a minor official in the Taliban Ministry of Culture and Information, he fought with the insurgents against the Americans. This information may or not be true. What we know for sure is that since 2007, someone named Zabihullah Mujahid routinely told the Taliban's side to Afghan journalists over cell phone calls and in text messages as well as emails, tweets, and postings on websites. Some journalists claimed he was more informative than the Afghan government. Still, none of that experience seems to have prepared him to be particularly good at his new job.

Although the Taliban social campaign did not add to much, it did get my attention. Here is why.

Zabihullah Mujahid just happened to be breaking news of the day, the day I decided to finally get started writing this book. It was more than happenstance. I needed a bit of inspiration. He delivered. Despite all the hype of social media prowess in the press, Zabihullah Mujahid offered a great exemplar of someone who got it wrong. This is something that can be explained by unlocking the keys to success in fighting with digital social networks online. So that is where *Digital Dominance* starts.

At the same time Zabihullah Mujahid peddled his "soft-gentler" Joseph Goebbels imitation of nice Nazis, another turn of events was influencing this book as well. A handful of Americans were delivering a masterclass in how to really win at social networking.

It all started days before Zabihullah Mujahid's press conference with an email from an old friend and mentor who served on the board of the Center for International Private Enterprise (CIPE), a nonprofit that works with individuals in the private sector around the world on free-market economic initiatives (more on CIPE in chapter 12). He wanted something. After the US invasion, CIPE had come to teach entrepreneurship in Afghan society. Now it was organizing a frantic effort to get people out, innocents who would most assuredly be targeted by the returning old regime. My friend's request was "help."

Afghanistan was half a world away in an opposite time zone, mired in absolute chaos and conflict. Helping from a computer in Washington, DC, seemed a forlorn hope. But when friends ask, you answer.

Riffling through my Microsoft Outlook contact list produced about a half-dozen email addresses of nongovernmental organization (NGO) representatives, former government officials, journalists, policy analysts, activists, and retired military—a wish list of folks whom I figured might know something, someone, or have a clue of whom to ask.

To be honest, there was zero expectation that this would amount to anything. At least, however, it was not doing nothing.

To my great surprise, the emails triggered a small hyperactive network, folks all seeking to work parts of the same problems—helping people, offering to help people, and people looking for help, efforts that spiraled into other networks of other people all rushing to save lives before it was too late.

In Afghanistan, the frenzied evacuation, reduced to the single airstrip at the commercial Kabul airport, surrounded by surging mobs and countless

Taliban checkpoints, proved a nightmare to navigate. A maelstrom of foreigners, Afghans who had worked with the United States and their families (and according to one report, a truck driver who wandered over to see what was going on and wound up on a flight to the United States) were trying to flee to the airport, scrambling for seats on any available plane.[4]

First, there was the issue of getting safely to the airport through the crowds and hordes of armed, bearded, scowling Taliban. That often required security, intermediaries, and bribes. Then there was the challenge of access through the gates. That meant having the right paperwork or more likely pressing to the front and convincing US troops for whatever rhyme or reason to let people pass. One of the most iconic and tragic photos of the time was a mother handing away her child over a razor wire fence into the arms of an American soldier.[5] For others, it was not unusual to sit outside the airport gates for a day or more without food or water, under the hot sun and the constant threat of beatings or harassment from the Taliban.

Even getting into the airport didn't mean getting out. It was chaos. Evacuees had to get manifested on a military or commercial flight. Then they had to arrive in a country that would accept them. If they were refugees, they needed food and shelter as well as everything from diapers to clothing, toothbrushes, medical care, money, and a place to live for starters. At the time, the US government was coordinating little of this.

As requests for help or offers of assistance popped in my inbox, they were consolidated and distributed to an expanding email chain. The list included a cadre of do-gooders from a former national security adviser and ambassadors to people who frankly I had no idea who they were (but they seemed like they knew what they were talking about, and at the time that seemed good enough). Often, I didn't know if I was emailing someone down the block or in Qatar, the United Arab Emirates, or Tajikistan. One of my new best friends became Terrell Chandler, the director of operations for Spirit of America. Terrell had served with US Special Forces teams in Afghanistan. She also helped train women who joined the Afghan army. Within days after the crisis kicked off, she was directing aid in theater in real time. More about her and her amazing team later.

I also posted on my LinkedIn account. That triggered several desperate calls for help and some new leads. That information got folded into our email network.

Nothing about this was normal. I even got a request from Fox News who was trying to find flights for reporters into Afghanistan—while everyone else was trying to get out!

After a couple of days of email updates, cutting and pasting a pile of requests, offers, and contact information, and coffee and more coffee, it seemed liked lots of people were connecting to lots of people. Not wanting to overload inboxes, on August 18 (the day before Zabihullah Mujahid's first press appearance) I offered to just back off and stop pestering people. I did not want to be a nuisance. Nothing is worse in a crisis than people butting in and kibitzing just to prove they are relevant. In crisis mode, it is crucial to pare down the network to those who make decisions and hand off critical information and not flood the conversation with distracting chatter (see chapter 5 on managing a crisis).

I didn't want to be a distraction.

So I thought it made sense for the person whom I thought had the least to contribute   me—to just back out. That triggered a flood of email requests to keep doing what we were doing. I relented, replying, "Okay, will keep spamming the group." And so we did.

Our ad hoc digital network continued to light up round the clock, tolerated by the good graces and understanding of my wife over emailing at dinner, during our favorite show, *America's Got Talent*, and late into the night. Since folks from almost every time zone in the world had something to offer, this turned into the network that never slept. More coffee. More coffee.

To be honest, it was kind of inspiring, watching all this goodwill and dedication unfolding on email. At one point, responding to a request for help, young volunteers on my staff spent their whole weekend consolidating lists of requests for clearance with the State Department. Who says Gen Z are slackers? This was the digital version of the British going down to the docks and across the channel to rescue the army at Dunkirk. All done without a Churchill.

Even noble efforts, however, come to an end.

On August 26, suicide bombers struck outside the airport perimeter killing scores including thirteen American military personnel. After that, it was almost impossible to get anyone out. The last US troops withdrew days later—on August 30.

There was still plenty of work to do: helping refugees already evacuated, trying to get others to safety over land. An underground railroad was emerging from a patchwork of NGOs and former Special Forces. There were also even more complicated situations, like hundreds of Afghan pilots and their families who had flown Afghan military aircraft to Uzbekistan. The Taliban wanted them back. The Uzbeks wanted them out. The State Department didn't want to answer the phone.[6] None of these were problems that were going to get solved in a New York minute. The value of our crisis group had run its course.

Still, there is no question some people got help who likely would now be dead if not for the chance connections of emails flying across the internet. There were also frustrating failures that played out in real time right in my inbox. Groups huddled, terrified and desperate, hiding out only ten minutes from the airport. There was one effort to evacuate hundreds of students (many of them women) who had attended the American University of Afghanistan. At one point, an email exchange secured open seats offered up by the US Institute of Peace who had been organizing flights to get their people out. An escort to the airport got some of the students past the Taliban checkpoints. Miraculously, they got to the north gate in time for the flight, but the gate never opened. The pilots had waited as long as they could, but faced with losing their takeoff clearance, the flight reluctantly departed without them. With their cover blown after trying to flee the country, the students and their families frantically dispersed to a series of makeshift safe houses. Later, chartered buses tried to drive them out, but after sitting at the Tajikistan border without permission to cross from the Tajik side, the Taliban demanded the buses return to Kabul. Weeks later, thousands of former students were still in Afghanistan at the mercy of the new regime.[7]

Nevertheless, this was powerful stuff. In less than a day, what became a global network of volunteers, many of whom didn't know each other before the crisis, effectively linked federal agencies, foreign governments, NGOs like Spirit of America (more on this group later), commercial charter plane companies, Special Forces veterans, a former yoga instructor, philanthropists, journalists, random do-gooders, a well-known writer of spy and political thrillers, and all kinds of other folks from all over the country, all over the world. This was hardly the only network either. They sprung up like mushrooms after the rain.

What these folks achieved was extraordinary. Although the collapse of Afghanistan, the resurgence of the Taliban, and haphazard withdrawal were a humiliating setback for the US government, if not for the incredible individual efforts of so many, from so many countries, assisting in the evacuation, the disaster could have been far worse.

There is a book to be written about all of this.

This is not that book. In other words, this is not a book about what they did. Rather, this is a book about why what they did worked.

Here is the question that started it all: What did my friends and colleagues get right that Zabihullah Mujahid got wrong? Here is the answer: They followed a formula for success that makes all the difference in building a robust, resilient, and effective social network in times of war, conflict, and competition, a network that can produce real results on the ground. That is what this book is about. *Digital Dominance* spells out the secret sauce.

Wiki at War

Electrons can get you killed or save your life. That was the message of *Wiki at War: Conflict in a Socially Networked World* (2011). At the time, digital social networks had already become all the rage. MySpace had already come and gone. Yahoo! was seventeen years old and well past its prime. Twitter was a youthful five, and the typical Twitter employee, if asked, would have responded, "Elon who?" Instagram was an infant at one. Meanwhile, on the dark web, the parts of the internet that can't be easily accessed with search engines and IP addresses, mysterious groups like the Silk Road (much more on these bad boys later) were building criminal empires. Today, emailing, texting, posting, and videoconferencing are more second nature to us than phone calls, writing, and reading newspapers were to the Greatest Generation. But while online social platforms had become ubiquitous, impactful, and rapidly proliferating in every field from shopping for shoes to national election campaigns and cybercrime, *Wiki at War* pondered how their use would play out in competitions that impacted national security in everything from terrorism to terrifying wars.

Here is what *Wiki at War* accomplished. Drawing from both scientific literature and practical examples about real people in the real world, *Wiki at War* fleshed out the structure and dynamics of how social networks functioned; why they were potentially powerful; how they could be applied to national security challenges for good or ill; and what distinguished casual networking from cyber superpowers.

At the time, there wasn't a lot of stuff in the public space about the application of social networks to national security. In 2014, I pitched the idea at the edgy and hip South by Southwest festival (the annual "Burning Man" of media, music, film, culture, and tech hoedown in Austin, Texas). I can't say I rocked the conference, let alone the world. I sold one copy at the book signing. That turned out to be to an old friend I hadn't seen in years.

Still, what I had to say mattered less than what happened. The forecasts of *Wiki at War* proved themselves in real life. Political leaders, protesters, armies and insurgents, terrorists, and the private sector all engaged in digital social media from "gray zone" competitions to espionage, disinformation campaigns, outright warfare, and humanitarian operations, often with dramatic impact.

Perhaps the most powerful contemporary expression of the disruptive power of social media was ISIS, the terrorist group that overran large parts of Iraq and Syria (2014). ISIS attracted not only global attention but also global support, drawing over thirty thousand foreign fighters from countries and walks of life all over the planet through a world-class social media campaign. Unlike Zabihullah Mujahid, ISIS wasn't trying to deceive the press about who they were. They wanted the world to know exactly who they were. That was why they executed a man by putting him in a cage, burning him alive, and posting the event on YouTube.[8] They were recruiting. They were advertising their war on the world.

Until they were stopped by the force of arms (2018), ISIS served as a posterchild for the potential of what could be accomplished online.

Just being right was not enough reason to write another book. What still needed to be explained was why one terrorist group mastered the art of Wiki war and another missed by more than a little. What made all the difference? Further, it was important to understand not just who dominated networks but also how that was done and how that power translated to influence. Armed with that knowledge, competitors would know how to win as well as how to frustrate the effort of their opponents. How to do all this was something not entirely clear when I wrote *Wiki at War* back in 2011. It is super clear to me now.

So, whereas *Wiki at War* described what made for an online winner, *Digital Dominance* instructs on how to become that online world champion and, conversely, how to take down your adversaries and leave them in cyber dust.

Digital Dominance unpacks the art and science of winning in four sections.

Part 1 covers how we got here from there and where we are now. The first chapter reviews and updates the ideas introduced in *Wiki at War*, describing how social networks work and what enables voices online to dominate a competition. The next chapter explains the secret to achieving dominating status, turning influence on digital social networks into action in the real world.

Part 2 of the book lays out the steps for building an effective network.

Part 3 details the steps for taking apart the networks of adversaries.

In part 4 of the book, the final chapters offer an overview of the contemporary digital environment where the war for dominance is playing out, surveying prominent state and nonstate actors, as well as the role of emerging new technologies.

Why Read On?

In *Wiki at War*, I recalled having read a history about insurgents in Malaysia during World War II. The book was called *The Jungle Is Neutral* (1949). The jungle didn't favor guerillas or invading armies. It was just jungle. It was just there. The jungle didn't pick a side.

Technology is the same. It doesn't take sides. People do. How people choose to use technology is what really matters. Digital social networks have unprecedented potential to engage, empower, and entertain or to spread despair, hate, and uncertainty—to harm or to help. What they will do is not a question of inherent good or evil. Dominance results from the skills of those who master networks—and through them, the real world. *Digital Dominance* gives you the tools to make a difference in this fight.

The consequences of this fight could be world changing. On the one hand, social networking done right could be as powerful a force for shaping the future as gunpowder, the printing press, electricity, and computers. On the other hand, not preparing for this fight is like looking up to see the invading hordes armed to the teeth approaching your shores and then going back to your vintage game of Super Mario Brothers.

Postscript

Since I started this book, a great deal has happened in the world to demonstrate how right I really am. In the following years, I and others stood up ad hoc networks to deal with the Russian invasion of Ukraine and the October 7, 2023, attacks by Hamas against innocent Israelis that prompted a protracted bloody conflict and humanitarian crisis. Concomitantly, we have seen global controversy over the role of social networking technologies in everyday life: massive barrages of charges and countercharges over misinformation, disinformation, and fake news. Meanwhile, new technologies like artificial intelligence began to completely reshape the digital space. We have seen superpowers, and not so superpowers, push their way to the forefront of the global information frontier like a thirsty biker gang heading for the bar.

The game, as they say, is afoot.

Backstory

1

THE STORY SO FAR

No one is more a master of wiki war than Sara Carter.

I knew Sara when. We once lunched at the Monocle, a DC institution (since 1960) across from the Dirksen Senate Office Building—less a place to see and be seen, than an easy walk from the Capitol for good food and a solicitous waitstaff that calls everyone "senator," remembers your name, and always has reserved your usual table at the ready.

Back in the day, Sara was a new reporter on the Washington beat working for another DC start-up, the *Washington Examiner*. I was writing a weekly column for the paper. Since we were both covering national security, we got together to compare notes and brainstorm ideas. The concept for *Wiki at War* was still bouncing back and forth somewhere in the back of my head. To be honest, at the time I was still struggling to figure out how to send emails on my BlackBerry without getting carpal tunnel syndrome. I had no intention of attempting a book to explain how people behave online. I also had no notion when I met Sara that I was sharing soup with someone who would become an apex predator in the world of online social platforms—exactly the kind of networking Napoleon that in the future I would forecast in *Wiki at War*.

Part of what I later learned is that there is no boot camp for achieving networking wonders better than life experience. Perhaps this is because networking is as much about understanding humans as it is about mastering technology.

Sara has had life experience and then some. Her growing-up years read something like if Graham Greene penned an exotic travel novel. Her father was a Marine, a veteran of two wars. Her mom was an exile from Castro's Cuba. Sara grew up in Saudi Arabia.

With degrees that had nothing to do with journalism, Sara naturally wound up as a journalist, compulsively drawn by the gravity that attracts many investigative reporters—an insatiable hunger to get the stories others couldn't—and preferably going in harm's way to do it.

She cut her teeth in California with reporting on secret cross-border cartel tunnels, cutthroat drug trafficking, and corrupt Mexican officials. Young, pretty, dark-haired, and equally fluent in Spanish and English, she seemed more like a college student headed to Cabo for spring break than a crusading journalist rooting out stories that earned others a bullet in the back of the head and a lonely, wrapped-in-plastic unmarked desert grave.

From Los Angles, one step ahead of the Mexican Mafias, she headed to DC, which turned into just another place to get her mail while she went into the danger zone searching for more stories. Sara spent months in South Asia. She wrote heartbreaking stories about impoverished women and children addicted to opium. She embedded with Afghan troops on the border with Pakistan. She traversed the tribal regions of Waziristan with the army hunting insurgents. She was shot at and mortared. Sara also made friendships and sources that made her one of the most well-informed journalists in the West about the affairs concerning the likes of the Taliban, ISIS, and al-Qaeda. More about these contacts in a bit. She also met a guy who would change her life more than a little.

In Washington, Sara made a major media presence. In addition to writing for papers including the *Examiner, Washington Times,* and *USA Today,* she was starting to pop up on television in places like Fox and CNN.

She married Martin Bailey in 2010. The army veteran had been serving as a contractor in Afghanistan. He had children from a previous marriage. They added a few more, assembling a total of six, a small platoon all their own.

In 2011, while on a mission, Martin was injured in a blast from a roadside bomb. He recovered but completely lost his sight. It was tough on the newlyweds, but undaunted, they proved a strong and resilient family that refused to be sidelined by setbacks.

In 2017, Sara joined Fox News as a full-time contributor. Today, she is as ubiquitous as a reporter can be, covering stories from the US southern border to the borderlands of Afghanistan.

Sara is also a networking superstar.

Employing the whole social media tool kit from podcasts to videos on Rumble and YouTube to Twitter (now X), Sara has scaled her influence far beyond periodic appearances on cable news and newspaper stories. Using multiple platforms, she builds sustained engagement with her growing tribe of followers.

Sara is also the Jedi Knight of the small places, intimate networks combining discrete groups engaged in serious conversations. Many of her breaking and award-winning stories started from her web of contacts, sharing information in real time that few outside their group had access to.

In 2020, Sara turned her passion for networking and investigative reporting into the Dark Wire (TheDarkWire.com), a nonprofit for training a cadre from Gen Z on how to go after stories underserved by the rest of the media, in particular targeting dark people in the world who exploit the innocent and defenseless. "We are in an ancient battle against a common enemy: hate," she recalled. "I have witnessed violent hate firsthand while covering the wars in Afghanistan, Iraq, and right here at home on the streets of America. Hate is un-American. It destroys the fabric of our nation and builds a path to a dangerous future. But there is still hope in fighting the trajectory of hate in America. We need to expose it and take action." I was even roped into part of her network, helping organize training for would-be investigative reporters on how to cover the Uyghur genocide in China.

When the Afghan retreat debacle kicked off, not surprisingly, Sara was in the middle of all of it. On one end, her stories were making national headlines and shaping the views of many Americans. On the other end, Sara was working behind the scenes to help save friends left behind. She was an important link in the information chain. In the midst of our networking frenzy, I got a query from another friend, Lisa Curtis (she returns in chapter 12). No one knows more about South Asia and is better connected than Lisa. We used to work together before she went to the White House where she served for almost four years as the principal adviser to the president in the National Security Council. Lisa called to ask for Sara's number. If Lisa needed to run something by Sara, you know it was serious. That kind of stuff happened all the time in the frantic days running up to the final pullout.

An independent spirit with incredible journalistic instincts, boundless courage, and limitless energy, on her own without a road map on how to do any of this, without publicists or high-powered backing from major broadcasters or big tech, Sara stormed her own beachhead in the social media space. Her campaign moved on forward from there, and she is taking more ground all the time. In 2025, her energy and expertise earned Carter the president's nomination to be the director of the Office of National Drug Control Policy—back in the battling cartel business.

Origins

Wiki at War explains why there are Sara Carters. This chapter explains *Wiki at War*. Here is a summary and update of all the key points in the first book.

For starters, the story of online action on the internet is rooted in the core of our humanity. Networks are the most human of institutions. They are foundational to our evolution as a species, programmed into our DNA. Digital platforms are just the latest version of humans plus networks, using computer technologies to extend our human ability to engage with one another in the same way garage mechanics use a wrench to increase the leverage of their arm strength.

The more they learn about evolution, the more modern bone detectives (paleontologists working with other types of scientists that study human physiology) suspect that the key development defining humans was the ability to walk upright, an innovation that came before anything else, a shift in our adaptive biology that really distinguishes us from other animals. And here is why they think our ancestors thought that was a good idea—dating.

A "real driver" in evolution, postulated Owen Lovejoy (not the Civil War abolitionist but the anthropologist and anatomist at Kent State University), "was a mating strategy: human ancestors took to two feet to become monogamous."[1] Mammals that walked on two feet could use their hands to forage, carrying food back to the nest to feed a mate and offspring. In this manner, bipedal motion facilitated the family unit, a small nuclear network.

Once our distant ancestors could support families, their capacity to nurture small children provided the means that allowed time for their offspring to grow bigger brains after birth. Unlike other predators, human children did not have to leap out of the womb and fight for survival until they were several years old. That was an important evolutionary strategy that offered humans a significant survival advantage.

An important measure of intelligence in species is called the encephalization quotient—the EQ, the size of the brain relative to the size of the body. Since the brain of human babies at birth is limited by the size of the head that can fit through the birth canal, evolution could only deliver bigger brains through engineering a species that could feed, protect, and nurture a newborn until their brains grew sufficiently large and complex enough to handle the more sophisticated tasks necessary for survival. Thus, through the network of the family, evolution provided a path for humans to develop an impressive EQ, a real competitive advantage over other mammals that might have been bigger, stronger, meaner, faster (at the start) but over time not smarter than the cagey, wily humans.

Going down this evolutionary path led humans to embrace networking behavior.[2] One international research team concluded that as humans evolved, they diverged "from the great apes in such a way led to the production of so much food that early humans had much more time to do other things, such as socialize. They further suggest that such socializing, combined with the organizational activities involved in obtaining food led to the development of larger brains and from there, other uniquely human attributes." The race for mammal domination was on and the humans had a big advantage.[3]

In humans, bigger brains enabled developing more networking skills. Prehistoric families teamed up with other families. Collectives of Ice Age kin created tribes that could pool hunting and foraging activities. That led to another evolutionary networking innovation—language—one that proved a superb social tool for all kinds of survival activities from defending the village against saber-toothed predators to hunting mastodons and wooing cave girls.[4]

Indeed, scientists speculate that the reason the gene expressions that enable language shutoff in young children after they learn their first language was an evolutionary benefit. If members of the tribe only knew the language of the tribe (locking in their limited social networking skills), they were more likely to stay with the group, keeping around another helpful set of hands to provide for the community, rather than the young drifting off to join up with a clan on the other side of the iceberg (for more, go to *Wiki at War*).

There is plenty of evidence to suggest that these scientific guesses about the evolutionary role of language are right. Take the example of Papua New Guinea. When Australian explorers reached the forbidden highlands in the 1930s, they discovered innumerable tribes unchanged since the Stone Age, all operating independently of one another in proximity with little or no interchange. This evolutionary legacy endures even today. In the island nation, Papua New Guineans speak over 850 languages.[5]

So here is what we know. We are designed by evolution to use networks. Our brains are equipped to use them as instruments, important for communication, productivity, and survival. Thus, it should come as no surprise that we have a hardwired attraction to technologies that can be used as social networking tools.

There is a school of thought that "technology develops independently from human activity, and that it is the primary driver of what happens in the human history."[6] The development of all communications technologies, not just digital social networking tools, argues exactly the opposite. Human impulses drive the adaption of technology.

The relationship between human choices and the development of communications technology is superbly demonstrated by Paul Starr in *The Creation of the Media* (2004). "Technology and economies cannot alone explain the system of communications we have inherited or the one we are creating," he writes. "Their development is impossible to understand without taking politics into account."[7] Politics, laws, rules, consumer choice, advertising, advocacy, lobbying, violence, the interplay of human choices, desires, preferences, happenstance, and needs shape when, where, and how we network—from the printing press to the internet. For instance, the adoption of the telephone in the United States significantly outpaced Europe because of different public policies.[8]

To sum up, our brains are programed to grasp at every emerging communication technology as it comes along and try to use it as a networking tool for our survival, our betterment, our entertainment, for power and pleasure and profit, to save our souls, sell shoelaces, or win votes. We did this with writing, the telegraph, the telephone, computers, the iPhone, and we will do the same with technologies that follow on from that. And like when we adopted all of those, we will make good, bad, indifferent, better, and worse choices. That is how the human operating system operates.

UPDATE

What was forecast over a decade ago has become every day in contemporary society. Digital social networks matter. Now let's bring the story up to date.

As the power of social media has become so readily apparent, it is no surprise that authoritarian regimes have invested so much effort in either suppressing or controlling the access of their citizens to the internet.[9] Even before cataclysmic events like the 2010 Arab Spring that sparked political unrest throughout the Greater Middle East, they recognized that online networking could be a tool to generate mass political action.[10] Events of the last two decades have demonstrated that influence time and again. The latest earthshaking instance of that occurred at the outbreak of conflict between Israel and Hamas; both sides sought to mobilize worldwide opinion in support of their cause, triggering demonstrations and calls to action on every single continent (Antarctica excepted).[11]

Democracies have struggled as much with governing in a socially networked world as repressive regimes have fought to hamstring the freedom of using digital space. In the United States, for instance, debates over the power of "big tech," arguments for regulatory reform or breaking up social media companies, fretting over "fake news" and "disinformation" have become pervasive and increasingly severe and vitriolic. We are all over the place in trying to address these concerns.[12] Some efforts to make the internet safer undermine free speech or economic freedom. Efforts to protect democracy might also promote privilege or prejudice. All we can seem to do for sure is to come up with solutions that we all do not agree with. More on these controversies and how they may affect networking online later.

There is no question that left unfettered by dictatorial regimes, democracies struggling to make sense of how to practice good governance in an online world, or a lack of access to the internet and digital devices, there remains a strong human compulsion to get online with others. This is also true despite a chorus of concern that online platforms are making users more antisocial.

There is some truth to the claim that being on social media doesn't necessarily add up to more social interaction. For starters, not everyone uses social media to be social. According to one survey, about 40 percent of users reported they primarily used social media to get news. Thirty-eight percent used the apps as entertainment platforms. Only 39 percent used apps for communication purposes.[13] These statistics, however, are a little misleading. Communicating in networks is not just about sending intentional signals to one another. Influence can emerge from all kinds of online activities from promoting news stories (such as occurs on Reddit) to comments on YouTube videos. All activity online that delivers content and connection, information exchange between humans, actions that could potentially shape views, attitude, and behavior is social networking.

Diversity in what people use networks for online is only one variable. Even where broad access to the internet isn't an issue, there is anything but a uniform distribution in the use of social media. Take the United States, for example. According to a survey by the Pew Research Center, almost 90 percent of Americans, age eighteen to twenty-nine, use social media sites. In contrast, of those sixty-five years and older, only about 45 percent frequent social media. Ethnicity appears to impact as well. About half of the Blacks and Hispanics surveyed used Instagram, compared to only 35 percent of whites. Other differences are even more pronounced. Forty-six percent of Hispanics surveyed frequented WhatsApp, compared to 16 percent of whites. Education matters too. Not surprisingly over half the college educated graduates responded they used LinkedIn, the career web-sharing site. Only 10 percent of high school graduates, however, reported signing up for the app. In addition to socioeconomic status, gender counts. There are more women than men on Pinterest, Facebook, and Instagram. There are more men than women on Reddit and Digg. Also, surveys suggest different groups have different preferences. Men are more interested in looking for information. Women are more interested in connecting with others.

In addition, there are all kinds of online tools that although not thought of as social media have become powerful tools for communication and networking. Many of these either didn't exist or were only modestly used a decade ago. From video networking within Teams and Zoom to new chatting features on X (with the promise from Elon Musk of a lot of other new innovations to come), the tool kit is only expanding, making online social networks more flexible and resilient.

It is not just disparity of usage that makes the internet a messy place. There is a lot of abhorrent behavior online from outright criminal activity like identity theft and fraud to cyberbullying, sexting (sharing sexually explicit material) and fraping (the combination of the terms "Facebook" and "raping"), defacing, commenting, or even taking control of another individual's online content.[14]

There is also a deep layer of online social networking that isn't measured by pollsters—the so-called dark web, the parts of the internet that cannot be easily accessed by most users. These are places where all kinds of malicious networking are ongoing, from scams and child pornography to murder for hire and terrorism. In addition, new forms of encryption provide new ways to shield online networking from the eyes of outsiders.

What is going on in the dark underbelly of the internet can be daunting— way more impactful than just the equivalent of cyber-pickpocketing. Take the example of Sina Weibo, a microblogging site that is one of China's largest social media platforms (more than twice the size of X). In 2020, Sina Weibo reported a data breach that compromised about 90 percent of its users—well over 500 million (more than the entire population of the United States), including all kinds of personal information from usernames to phone numbers. Reportedly, the attacker sold the whole database on the dark web for the equivalent of about $250.

Every aspect and individual online is potentially a victim of cyber bad behavior. A 2023 report, for instance, concluded that US youth are increasingly a target for cyber scams, losing a record $210 million to cybercriminals, representing a 2,500 percent increase in five years.[15] Cybercrime and malicious activity, a lot of it leveraging online social networking, is not just a problem for heavily cybered societies like China, the United States, and Europe. In a 2021 case, the African Network Information Centre (AFRINIC), a nonprofit that assigns internet addresses on the continent, sued a Chinese businessman to recover 6.2 million African internet addresses (about 5 percent of the total in Africa) that had been exploited for a range of criminal activity, from pornography to illegal gambling.[16]

More on the criminal and malicious side of getting together online later, but it's important to remember the shadow world of networking is just as crucial to understanding the digital social activity as the stuff that is above board. It is a digital jungle out there.

And it is not just the intentional evildoers we have to worry about. There is a vigorous debate over whether social media builds up the social fabric of society or undermines it. For instance, is online social behavior antisocial? Any parent who sat at the dinner table, with a cadre of kids that never look up from their iPhone or utter a word, might say yes! But beyond ignoring the humans around them, social networking activity has been cited as a threat to social cohesion inciting everything from depression to aggression.

Does a socially networked world really impede social networking? In 2018, two researchers, Samuel P. L. Veissière and Moriah Stendel, one a psychiatrist and the other an anthropologist, argued that social media doesn't make people antisocial; rather, it makes them "hypersocial." They concluded that we are impulsively drawn to the technology, "driven by the human urge to connect with people, and the related necessity to be seen, heard, thought about, guided, and monitored by others, that reaches deep in our social brains and far in our evolutionary past."[17] All that happens is that when we take our social actions online, we take all our human baggage with us.

Social networks reflect all the foibles, strengths, and diversity of real-world activities. For instance, mental health issues that might be triggered or exacerbated by online activity are not dissimilar from those that occur in everyday life sparked by everything from abusive relationships to substance abuse and genetic disorders. Rather than being a technologically hygienic safe space, the online world carries all the same peculiarities and pathologies of the material world. So, for example, some people who gamble online get addicted to gambling, just like when gamblers get hooked on poker at card tables in Las Vegas. In the real world and virtual space, humans act like humans. Big surprise.

The fact that the internet is messy is a reminder that understanding social media has more to do with understanding humans than mastering technology. In short, online social networks operate like all human networks reflecting all the aspects of our humanity from social and cultural differences and preferences to the altered states of our mental states. Networks are us.

Finally, we know what we do online appears here to stay. Twenty years of networking have shown that the appeal of social media is enduring and resilient. On average, today, every day in the United States, individuals spend about 2.5 hours online. That number (even when Americans sat home during the COVID-19 outbreak) remains stable. Where we go online has also remained consistent. Facebook and YouTube continue to dominate as they have for a while. In 2021, Facebook had 2.89 billion users (90 percent of them, by the way, outside the United States). YouTube had about 2 billion (about 80 percent outside the United States). What these numbers suggest is that online networking has staying power. It'll be around for a while, and here is why that is important.

Meet the Power Curve

What makes social networks particularly powerful? To answer that question, *Wiki at War* drew on an important insight from the technology writer Clay Shirky. In his book *Here Comes Everybody: The Power of Organizing Without Organizations* (2008), Shirky highlighted the distinguishing feature of digital social networks—that they conform to the power curve, a mathematical distribution that describes the functional relationship between two quantities.

In a power curve (also called the power law), the relative change in one quantity drives a proportional relative change in another. At the high end of the curve, this results in a dramatically lopsided asymmetrical distribution of values. At the other end, the curve flatlines with the quantities being virtually equal. The way the curve scales is dramatic, dropping like the slope of Bob Hope's nose. Think the letter *U* cut in half and turned on its side.

In the real world, this breaks down in what is often called the 80/20 rule, as in 80 percent of the work is done by the 20 percent top workers, or 80 percent of the complaints come from the 20 percent "Karen" customers. In a digital social network, the two variables are (1) the number of participants in the network and (2) how much activity they do online. That pretty much means 20 percent of the people do 80 percent of the texting, emailing, posting, and commenting.

Here is a simple example of what that looks like in practice. I have a global listserv of people I connect with, including politicians, researchers, journalists, and activists. I do almost all the emailing. Rarely does anyone else contribute anything. Yet, it is a real network. Here is how I know. Folks who work with me travel all over the world. Invariably, they will come back and say, "Hey, I ran into someone, and they asked if I knew you. They love the stuff you send out." The point is, I may be doing most of the digital talking, but a lot of folks are digital listening.

The power curve has two significant implications for the conduct of online activity. The first one has to do with scale. Growth can be exponential. When networks start to scale, they become massive, a Godzilla of online reach. In short, a social network can offer the potential to engage a lot of people. Recently, the most followed person on Facebook, for example, is the soccer (football) player Cristiano Ronaldo with over 151 million followers.

When Twitter (now X) started in 2006, the app had 140 users. By 2021, it had over 200 million. That's scaling. At its height, the Roman Empire probably had only about 120 million people (and that's the highest estimate). If Diocletian could have contacted them all with the click of a mouse, we would probably all still be speaking in Latin. Instead, he found the empire had grown too large and unwieldy. It was impossible to connect effectively. He split it in two. The barbarians came. End of story (though, to be fair, the eastern half of the empire lasted another thousand years). The point is, both size and reach matter. Unlike the Roman emperors, Facebook influencers and TikTok stars can do what Rome could not—achieve both size and reach.

The second implication has to do with influence. As the curve scales, the bulk of the communicating is done by fewer and fewer people. At the apex, a few in the network (following the 80/20 rule) are doing most of the talking, and influencing, to the many. Here, the curve operates in what I called "broadcast mode."

Take the example of Instagram, the photo-sharing app. About one billion people follow other people on Instagram. In 2021, I had 567 followers (the mean number of followers is about 150, making me above average). In comparison, where the curve really starts to scale, only 0.94 percent of users have more than fifty thousand followers. About eighty accounts have over a million. At the pinnacle of that elite of 2021, Addison Rae had a whopping 38.7 million users following her beauty tip posts.

The 80/20 rule explains the power of people like Sara Carter. When they grow big social networks, they become more than just global kibitzers; they become global influencers.

One way to become a Napoleon of digital platforms is to dominate a large social network. Platforms like YouTube, for example, figured out the power of the way the power curve works—and figured out how to make money off it. YouTube pays contributors on the platform if they amass at least one thousand subscribers to their YouTube channel. The more subscribers, the more they get paid because YouTube knows the bigger the channel, the more its potential to influence more people—and YouTube figured out how to make money off it.

In 2020, YouTube's top earner was Ryan Kaji, scooping up $29.5 million in profit for his 41.7 million subscribers. Ryan was nine years old. He makes short videos for other kids along with his parents and twin baby sisters.

In turn, YouTube makes money by selling ad space to accompany Ryan's videos. Using the power of the power curve and dominating influencers like Ryan Kaji, in 2020, YouTube amassed a total of almost $20 billion in advertising revenue.[18]

The other way to exploit the dynamics of social networks is at the flat end of the curve. When the network gets so small, all the members of the network tend to contribute equally. This is "conversation" mode. At this end of the curve, what makes for a Zen master is the measure of the quality of the participants and the information they share that are brought into the network. Not surprisingly, a group of four chatting Nobel Prize laureates produces a different quality of dialogue and decision-making than four preschoolers on the monkey bars debating the virtues of SpongeBob SquarePants. Rather than the quantity of the network reflecting the power to influence, in small social networks it is the quality of the information exchanged or created that makes the network valuable.

Take the example of our Afghan crisis response network. At the height of our activities, there were probably never more than two dozen people in the email chain.

Sara Carter is an exemplar here too of the power of what a few can do. As late as the day of the bombing at the Kabul airport, she was using her personal network to help get people out of the country. "We were able to get several people in yesterday and one this morning just before the bombings," she wrote me later that night. "But a young woman . . . had to do it on her own. . . . They told her that she was not gonna be able to pass and I told her not to leave and then an angel . . . snuck her in. . . . Her plane was just leaving the ground as the bombs went off." This small miracle was a testament to the big power of small networks.

The Art of Wiki War

Social networks operating in "broadcast" or "conversation" mode both offer a distinct edge for dominating activities online. But what is really intriguing and powerful, what really maximizes the exploitation of social networks is to have the capacity to link these complementary advantages together.

Broadcast mode is more useful if there are important and impactful things to communicate to a lot of people. Broadcast mode also allows for surveying vast amounts of chatter, gaining situational awareness within the social networking space, looking for crucial bits of information, signs or signals that could offer a competitive knowledge advantage.

Conversation mode is an effective way to analyze, assess, and develop knowledge with a highly qualified team. Information, for example, can be transferred from a big network to a small high-quality network where it might become the kernel for developing a new insight or competitive advantage. This is like strip mining ore in the information world and then smelting it down into information gold.

The two modes can be linked in a synergistic manner. Key units of information can be shuttled back and forth by key influencers alternating between broadcast and conversation mode, establishing a cycle for outthinking, outtalking, and outacting an adversary.

In the crucial days running up to the closing of the Kabul airport, the broadcast/sharing information dynamic dominated how our networks worked. Folks like Sara shared real-time updates that could be blasted out to others, everything from what gates at the airport were open to security threats to watch out for. In fact, days before the bombing at the airport, we were circulating warnings about possible plans to stage violent acts outside the perimeter.

Identifying the high ground in social networks, whether at the top of broadcast mode or in the middle of conversation mode, is one thing; taking the high ground is another. Exactly how humans engage in online networks, why some videos attract, and some emails are ignored reminds us that these are at their core human networks driven by human action, desires, and initiatives, not technology. Action and reaction in digital networks are social, not mechanical. This was a topic explored in some depth in *Wiki at War*.

What laws of social behavior are most strongly manifested online? One factor the book looked at was the dynamic popularized in the 2004 book by *New Yorker* columnist James Surowiecki, the "wisdom of crowds," which holds that the median judgments of large numbers of people drawing independent conclusions are consistently more accurate than a single individual or even a group of experts. That might seem like a powerful force in large networks. In practice, however, that is not what happened.

Here is the reason why. Networks are not just a collection of independent actors. Being part of the network, the participants in the network are influenced by the network. Influencers can push us in one direction, or they can work at odds with one another, skewing attitudes or frustrating consensus building. Benjamin Golub and Matthew Jackson, for instance, described two "obstructions" in online collaboration, when there are "extreme imbalances in trust" or a "lack of dispersion" when small groups or individuals stubbornly refuse to pay attention to input from the rest of the network.[19] In short, there can be all kinds of subtle and not-so-subtle voices calling us to one view or another. Such frictions make crowds sometimes look more like lemmings than the three wise men.

In the past few years, we have seen how some online behaviors and activities have demonstrated powerful, sometimes commendable, and sometimes unhealthy outcomes. One of the more notable cautionary tales involved Facebook, which owns the popular picture and short video–sharing site Instagram. In March 2020, Facebook announced it was going to launch its Instagram Kids app. A few months later, reporting in the *Wall Street Journal* claimed that Facebook had made this decision despite being aware of extensive research concluding Instagram had contributed to mental health issues in teenage women from depression to eating disorders. In September, Facebook announced it suspended the project. The company had based its initial decision to expand in the preteen market on extensive research on how the product could be used to engage and interact with children, which apparently also revealed the potential for mental health side effects of protracted use.[20] The site then cautiously retooled its program, requiring users be at least thirteen years old and introducing a host of parental control tools.[21]

Behavioral manipulation by networks can trigger good impulses as well as bad. Giving online has become an increasingly dominant part of the philanthropy space. In the United States and Canada alone, for instance, in 2020, crowdfunding (raising money from many small donors for projects including philanthropy) raised almost $80 billion. One of the most successful sites is GoFundMe, a US-based platform that allows individuals to raise money for philanthropic causes without having to set up a formal nonprofit organization or charity.

Social activism is also prevalent online. This was already a thing in 2011, but the growth since has been exponential. These days, there are social activism campaigns in virtually every corner of the world. Even movements in regions with limited internet access and scant broadband coverage have an online social media component.

Some of the online social activist campaigns have had global reach. Who can forget, for example, the famous 2014 Ice Bucket Challenge, where folks posted videos and pictures online of ice water being dumped on people to raise awareness and money for research on amyotrophic lateral sclerosis (ALS)? The craze swept up participants from former US presidents and the pop star Lady Gaga to one-sixth of the entire population of Great Britain. Combining social activism and philanthropy, according to one assessment of social media activity, resulted in the posting of over 17 million videos. All that ice water produced results. In the United States alone, the campaign raised $115 million that year.[22]

Philanthropic online giving and social activism have become a global phenomenon. Like many aspects on digital social network, in practice the activities reflect the political, cultural, and social norms of users. For instance, despite the rigorous internet controls in China, social activism and digital philanthropy are very active. "The activists tend to focus on areas that the government is not overtly against and may even support in principle but hasn't prioritized yet," Michael Cunningham, an expert on Chinese political affairs, told me. "In that way, they prod the process along an inch at a time. But it's really dangerous—they have to try to drive awareness of their cause without getting so much attention that the local or central authorities see them as a threat to public order." It is an endless game of online cat and mouse. The Chinese government and the Chinese Communist Party invest an enormous amount of effort speaking to their people online as well as monitoring what all those people are doing. No wonder China has 1.3 billion people on the internet—the most in the world. How this plays out and continues to change and evolve is one of the most interesting dynamics in the online world.[23]

Like every other aspect of online action, even the upside of doing good on the internet has a downside. For example, all that online action doesn't always equate to, well, action. Take the case of the "Save Darfur" campaign, highlighting an ongoing genocide in Africa. According to one study that looked at 1.2 million users who signed up for the movement on Facebook, most participants recruited no one (other than themselves) and garnered very few financial donations. What emerged was a massive online social movement that had virtually no impact in the real world.[24] Sometimes, the internet is all talk and no hat.

Online social activism can also be used for campaigns for questionable public "goods." Recruiting and proselytizing efforts by the terrorist group ISIS are one example.[25] Activism can also create risks for activists. In 2017, for instance, Julie Uldam published research findings that the energy company BP used social media activity to surveil activists critical of their policies and practices.[26]

Philanthropy online is another case in point. Online philanthropy faces the same challenges as philanthropy offline. Not everyone operates with good intent. In 2017, a woman told a heartwarming story of how she ran out of gas and a homeless man spent his last $20 to help her out. She started a GoFundMe campaign to help get him off the streets. Their goal was $10,000. The campaign raised $400,000. The only problem? It was all a scam. They both wound up in prison.[27]

Many of these activities have little to do with national security per se, but the dynamics of how they work are the same that govern national security networking—because they are all human activities. Therefore, those looking to understand how the dynamics of national security networks playout can learn from studying all kinds of networks.

Furthermore, virtually any network can be knowingly or unknowingly enlisted in national security–related actions from cyber-spying to devilishly spreading disinformation. A 2022 report by the Organisation for Economic Co-operation and Development (OECD) concluded, "Russia's war of aggression against Ukraine is notable for the extent to which it is being waged and shared online."[28] In global competition, all networks can matter—maybe.

The bottom line is that a decade after writing *Wiki at War*, networks are having the dramatic impacts predicted but not always. There are squirrelly things going on when the real world and the online world connect. That is a topic that needs to be unpacked further.

Wiki People

What makes people, from hucksters on GoFundMe to innovative governments, good at this? In retrospect, I started out with some good ideas that turned out to be more right than wrong.

The Afghanistan crisis offered many examples of both effective online warriors and digital blundering. We saw similar outcomes in the aftermath of the Russian invasion of Ukraine in February 2022, where Ukraine appeared as adept or even better than the Russian online behemoth at delivering outcomes through social media.[29] Watching many practitioners in action, I was right when I wrote *Wiki at War* about what kinds of people (like Sara) would be particularly good at these activities. I thought the best approach was to explain and explore the two ways humans work and how their actions and interactions impact digital networking.

Humans basically organize activities in two different ways—linear and nonlinear.

Linear just means proceeding in doing or solving anything by following a sequence of activities or actions, like going from a to b to c. Linear processing is awesome for activities where the steps and the outcomes are predictable. Thus, we have bureaucracies. Bureaucracies are built to conduct routinized activities day in and day out.

Governments are mostly bureaucracies. Sometimes, we like that. There is nothing wrong with being organized, efficient, and dependable. When the US government delivers Social Security checks like clockwork month in and month out, the recipients appreciate that they can depend on that.

But government bureaucracies are also not very agile and can prove maddeningly unresponsive to dramatic change or unanticipated activities. It is not surprising to see governments like Egypt, Iran, and Cuba when faced with mass action online respond by being little nimbler than using closed fist and just shutting down access to the whole internet. The Russians have a well-earned reputation for being devious online, but a lot of their malicious cyber activity looks like endless chaotic, repetitive action, a "throwing spaghetti at the wall and hoping something sticks" approach. Some of the stuff governments do online looks so amateur that it gives amateurs a bad name.

Even the best of governments can also be clumsy online. Look at the United States, the home base for some of the most innovative tech companies in the world. Then look at federal government websites and explain how that happens. When email started to take off, the US government "stretch" goal was to get 10 percent of the federal workforce on email. In retrospect, that kind of initiative seems laughable. That should have been a bright red flag highlighting that traditional governments are too sluggish to keep up with the rapid shifts in technology and user preferences that rule the social networking world.

Furthermore, it is not just government. None of the leaders of the social networking world were the dominant technology companies around at the birth of the digital revolution. Where is IBM's version of Instagram? Where is General Electric's platform for Gen Z? Heck, where is Blockbuster (at the time the world leader in marketing video movies and gaming)? Actually, we know the answer to that one. The last store is in Bend, Oregon. The point is, major corporations, which had brilliant boards and boundless bureaucracies all their own, didn't prove that nimble either. Something was missing, and it wasn't just shortsighted corporate judgment. Indeed, many of the companies that did emerge to become powerful players in the digital space prided themselves on having different corporate cultures that transcended the ways traditional companies operated.

The opposite of linear thinking is, obviously, nonlinear. Linear decision-making irresistibly moves in one direction. If the task at hand is to produce government checks, then choices revolve around getting checks out the door. Someone may find a faster or more efficient way, but in the end, they're cutting checks. Nonlinear processing is open to the possibility of going in different directions, of combining information in different ways, operating in a more intuitive and less structured manner. In the social networking world, it brings the art to the art of wiki warring. Innovation and intuition are particularly well suited to the less structured world of networking activities. Brainstorming is a good example.

Brainstorming doesn't mean what folks think it does. The image that comes to mind is a bunch of millennials around a whiteboard, sipping mocha lattes and tossing out ideas for a new social media campaign to replace plastic straws. Research suggests that way of creating innovation doesn't necessarily deliver the mind-blowing ideas we think it does. One survey concluded that the same number of individuals working alone produced more and better ideas than groups brainstorming together. The reason for that has to do with the friction that often results when unstructured groups start to throw around options.[30] That is not to dismiss the value of creative, nonlinear, intuitive thinking or the value of group interaction in honing and developing ideas. Indeed, real imaginative collaboration unfettered by linear structures produces all kinds of magical thinking that can lead to real innovation.

Transcending linear thinking can be a real competitive advantage. While the inventors of email just thought they were creating an electronic letter and the guys who started the World Wide Web believed it would be neat to have an electronic bulletin board, it was the innovators and creators who turned these ideas into iPhones and Outlook.

Still, awesome nonlinear thinkers have their limits. The fact is, the people with the best ideas also have the worst ideas because they have the most ideas. Furthermore, some brilliant imaginers don't necessarily have the organizational skills to get through a lunch menu. Sometimes, a nonlinear approach can be disruptive in a bad way, like the guy who thinks he can put IKEA furniture together faster and better by ignoring the instructions and just winging it.

When it comes to networking, both ways of thinking bring something valuable to the table. Organizational structure brings efficiency, discipline, and focus. Imagination adds intuitive judgment. Pairing the skills to get things done with the power to have great ideas is a powerful combination. Putting people with these talent sets in digital networks is like plopping Kristin Chenoweth on a Broadway stage. Magic happens.

Update

Since the publication of *Wiki at War*, we have seen a plethora of examples where leveraging both sides of the brain delivers better outcomes. When Walter Isaacson's biography of Steve Jobs came out (2011), for me, it was a revelation. Jobs's genius, his intuitive aesthetic choices and insights, was the spark for a tsunami of groundbreaking products. Apple really took off when the company figured out how to pair his wunderkind skills with a management structure that could build and deliver world-class products and services.

Not everyone is a Steve Jobs. There are lots of one-hit wonders we can point to online. In 2004, Gary William Brolsma achieved almost-instant global fame after he posted his "Numa Numa" dance video. In three months, on one site, the video had 160 million views. He got mainstream media coverage from *Good Morning America* to the *Tonight Show*.[31] And then he pretty much vanished. In 2019, he released an album, *The Haunted House of Pancakes*. You can pretty much guess how his career ended up.

Unlike the "Numa Numa" guy (no offense, Numa Numa guy), when a strong set of linear and nonlinear skills are found in tandem, online practitioners prove to be consistent and resilient winners and dominators. I saw this at play in our Afghan response network, particularly in the folks I had known and worked with for years, including Sara Carter. They were no slackers. In their careers, I had seen them pair their life experiences, creative talents, and organizational skills to deliver results. I was hardly surprised they delivered like networking all-stars in the face of the Afghan crisis.

Even governments can do this. Estonia is credited with having the most digitized governance in the world. Virtually every government service and activity, including voting, is online. To be fair, they started from a good place. After exiting the Soviet Union in 1991, they were saddled with some of the most pathetic infrastructure on the planet. They pretty much scrapped everything and began anew with the good fortune of starting at a time when digital services began to reach a level of maturity that they could deliver real capabilities. Estonia also had the advantage of being a very small country, with a small government and relatively trusting and cooperative society and blessed by strong collaboration between the public and private sectors, led by some smart people making smart choices. So once the old Soviet bureaucracy was swept away, there was less friction than in countries with overwhelming ranks of well-organized and regulated unions and federal employees to navigate. In short, Estonia achieved what it did by not acting like a typical government.[32]

Still, even Estonia has it challenges.[33] In 2020, it was bested by Denmark and South Korea in the United Nations E-Government Development Index, and there are several areas where other countries deliver electronic public services with more advanced technologies. Just like companies like Apple, terrorist groups, hacktivists, virtual activists, philanthropists, and others online, the Estonians have their ups and downs, but they have kept the skills that deliver a competitive edge. Thus, they remain strong players in the social networking world.

Tools for National Security

Looking at all these activities and their impact had a purpose bigger than just how to make money, influence friends, or serve citizens. The driving purpose of *Wiki at War* was grasping how all these were going to affect the fate of nations and the peace, prosperity, and security of their peoples.

The bottom line of *Wiki at War* was a warning. I love movies (spoiler alert—more obscure and seemingly irrelevant film references ahead). All-time favorite is the original *The Thing* (1951), especially the end—the "warning." Scotty the reporter menacingly announces how to deal with space aliens (the monster was played by James Arness, also the fictional Marshal Matt Dillon in the long-running TV show *Gunsmoke*). "I bring you a warning," Scotty proclaims. "Every one of you listening to my voice, tell this to everybody wherever they are." I wish I could have thought of lines like that. Maybe more people would have taken notice of *Wiki at War*.

Basically, what I concluded was that the uses of social networks for national security–related challenges was limited only by human imagination and the current state of technology. I thought that was even scarier than an alien invasion.

Furthermore, I argued that no player had an inherent advantage, that state and nonstate actors would fall on these tools with a voracious appetite like zombies chasing down a lame guy leaving a blood trail. This was not like the major leagues where only all-stars could play in the all-star game. While great powers like the United States, China, and Russia would no doubt jump into the fray, so too would states with more modest means from Israel to North Korea, as well as nonstate players from terrorists to hacktivists. Indeed, weaker actors would be more motivated to leverage social networking to offset the advantages of competitors with more money, power, and authority.

I remember once asking an expert on cyber operations what the United States was capable of doing online. This was pre–Edward Snowden (Snowden in 2013 leaked highly classified data held by the National Security Agency, the top US intelligence agency in the cyber world). My friend the spook, of course, said, "I can't tell you. I can tell you we can do stuff no one else in the world can do. I can also tell you, we can't defend against the stuff we can do." That pretty much summed it up. It was going to be a cyber free-for-all, at least for a while.

UPDATE

I don't have to catalog the range of national security activities impacted by social networking over the last decade. P. W. Singer and Emerson T. Brooking in their book *LikeWar: The Weaponization of Social Media* (2019) did a great job. They authored a dense book with lots of examples and stories if you are interested. For me, unfortunately that book is a bit like a cyber salad bar. Instead, let's boil it down to the key developments. There are four.

First, social networks as a weapon can impact national security, sometimes in very unexpected ways. Like the guy who tried to warn the scientists in *The Thing*—Don't try to make nice with the big angry alien; there are dangers out there. Take the example of the infamous Vevo hack. Vevo is a music video streaming service. Today, it has about 26 billion views per month. That is billion with a "b," including everyone from Ariana Grande live to Garth Brooks. In 2017, the company reported a massive hack of sensitive data. A little investigating revealed that the hackers snared a Vevo employee in a phishing attack (using a fake posting online to get someone to reveal sensitive information like passwords or to insert malicious software into their digital device that could then be used to hack into any system they logged into) on the professional networking site LinkedIn. The attackers were Russian operatives who managed to get not just into Vevo but also found dupes trolling for jobs on LinkedIn that helped them hack into the Department of Defense.[34]

Second, technology is changing the name of the game all the time. In *Wiki at War*, I warned social networking in the cyber world was not a challenge to be mastered but an environment to survive in. Technology is always a wild card. Players are invulnerable one day and compromised the next. The only constant is change.

Since the book came out, the innovators have not disappointed. They keep on innovating. One example is "deepfakes," fake or manipulated video, audio, or photographic content (often created with simple and widely available software tools like FakeApp and DeepFaceLab) that can be proliferated through online social media. A 2020 report from the Australian Strategic Policy Institute described several ways that deepfakes are being used in the national security space, including fake profiles generated on LinkedIn as part of suspected intelligence-gathering operations.[35]

The dynamic of technological innovation is one of the consistent aspects of social media competition that we can count on. And there is more to come. Two of the most disruptive innovations, quantum computing and artificial intelligence, are likely to turn the whole cyber world on its head. More on that later as well (chapter 14).

Third, it is a real competition. Sure, the big guys are fighting hard to stay on top. For example, the US State Department has an Office of Global Social Media that has a presence on all the major social media platforms, from

Facebook to Flickr. They text and post in foreign languages as well. There is a North Atlantic Treaty Organization Strategic Communications Centre of Excellence that looks at all kinds of social media activity, like the manipulated use of Russian-language media across Europe.[36] And it is not just governments. All major social media platforms invest significant resources in attempting to prevent the malicious use of their networks. Facebook, for instance, employs hundreds of people in centers tasked with combating terrorism online (we will meet one of them in a later chapter).

The little guys are busy too.

We know the great powers worry about other players online. Look at the investments they make to keep them in line. Over the last two decades, for instance, the Chinese government has invested extraordinary resources in trying to keep their own population under control. China's Great Firewall is a massive denial-of-service filtering system to block everything the government doesn't want their people to see. One study, measuring internet traffic flows over a nine-month period, tracked how enormous the effort is as well as how it works. They tested an average of over four hundred million domains a day. Of that number, the Great Firewall blocked three hundred thousand per day. In addition to sites like Facebook, the research also found on average the wall blocked forty-one thousand innocuous domains—in other words, sites that the Chinese government didn't consider an actual threat. Overcensoring is a serious inefficiency. For example, during the height of the COVID-19 pandemic, blocking inadvertently prevented students from taking online classes from overseas universities. Yet, it's a burden the government is willing to bear.[37] There is a reason for that. The Chinese government is worried that somewhere among the billion-plus people behind the wall, there are some who might cause them trouble online.

Fourth, keeping score doesn't tell you who is winning. As we noted before, the level of online activity doesn't necessarily indicate the level of real-world impact. ISIS, the Islamist terror group, achieved unprecedented global reach in its efforts to recruit foreign fighters. Yet, although their online voice was heard around the world, their influence was not. India, for example, has one of the planet's largest Islamist populations, yet online radicalization brought few recruits to the ISIS ranks. In contrast, Trinidad and Tobago, the tiny twin-island republic in the Caribbean, had one of the highest per-capita contributions among ISIS foreign fighters in the world. Why the difference? Clearly, there are other factors at play, factors that must be understood to comprehend how social networking impacts national security.

Where Do We Go from Here?

To summarize, this was the case made in 2011 and where we stand today. Networks are important to humans. Digital networks are so desirable, not because of new technology but because they help fulfill a fundamental need. Networking is a basic human tool to survive and thrive. Our brains are wired to use networks as tools to provide situational awareness, information, ideas, warnings, coordination, and validation (as well as cat pictures and mean memes). They are us. That is why we gravitate to them. And look, we are using them. Well over half the planet uses digital networks.

Furthermore, what we see is that when humans engage in social networks, they act like, well, humans. The same activities, attitudes, actions, and pathologies that we see in the real world from compassion to crime pop up in the online space.

Wiki at War also drew on the scientific literature to describe why digital networks were so potentially powerful. Following the power law, they have the capacity to scale, delivering tremendous reach with the ability of few to influence the many (broadcast mode). Small networks (conversation mode) can host high-quality engagements. Linking commanding heights of the two ends of the curve together provides a potentially dominating advantage.

At the same time, we have seen these networks deliver dramatic outcomes, not just online but also in everyday life, from places where people compete to where they try to cooperate, in corporate boardrooms, farming villages, across deserts, ice fields, battlefields, and ball fields. These activities in and of themselves offer insights into how networks can be used in addressing national security challenges. In addition, even the most benign networks can themselves be weaponized and recruited into the struggle between different entities online.

Wiki at War also talked about the skills and attributes of network dominators, forecasting that those equally skilled in linear organizational skills and nonlinear, improvisational, creative thinking would be the best at bringing out the best online. And indeed, we have seen many examples of people, organizations, and governments acting in just that way.

Finally, there is no question that social networks are tools with important national security implications. There are significant dynamics at play impacting how they work their mayhem on nations and nonstate actors alike.

That all leads to one question that must be answered: How do they do what they do? Answering that puzzle starts with the next chapter.

2

HOW TO WIN

David Shedd is (was) a spy.

I met David at a little Italian restaurant on E Street in southeast Washington, DC. The place had the feel of one of those nondescript joints where spies recruit spies, passwords are passed, or microdots were once slipped under the table. This wasn't too far-fetched a feeling. DC is a city of spies. Restaurants were, and still are, a common haunt. A Russian spy once evaded his American handler by walking out of a little French bistro (now long gone) on Wisconsin Avenue in Georgetown and strolling back to the Russian embassy. History can happen in a DC restaurant.

Shedd didn't look like a spy. He looked like everybody else, which is to say, exactly how most real spies look. He looked like gray-haired Uncle David.

There is another way to tell he was a spy. Look at his social media. There is none. Go ahead. Do an internet search. The stuff that is all public record, official biographical information and his work on some boards, and so on— that you can find. Other than that, he is invisible online. Don't mistake, the intelligence community is really interested in networking and social media; they just aren't social on the internet.

As it turns out, our meeting was just social. This was just lunch. David went to the same church as one of my colleagues who thought I would find meeting David interesting. It was. We then had another lunch—same little restaurant, same so-so pasta.

The next time we broke bread, Shedd mentioned he was retiring soon. That triggered a thought: "Hey, maybe he might consider thinking about affiliating with a think tank." Did I mention I managed research at a think tank?

If you are wondering why a think tank might be interested in David Shedd, it might help to know who he is. Shedd retired as acting director of the Defense Intelligence Agency (DIA). The DIA is the intelligence arm of the Pentagon.

Before heading the DIA, David had a long career as an intelligence operative in the Central Intelligence Agency (CIA). His service spanned decades going back to the Cold War. He worked in the field, mainly out of embassies in Latin America. He also served in a lot of high-profile jobs in the intelligence community, including as an intelligence officer on the National Security Council staff, CIA headquarters, and as the chief of staff for the director of national intelligence, the principal intelligence adviser to the president who also oversees the whole intelligence community (the family of organizations and agencies in the federal of government, including the DIA, CIA, Federal Bureau of Investigation, and National Security Agency, that collect and analyze intelligence). Shedd is a walking "who is who and what is what" of the US intelligence enterprise as well as the inner circles of friendly and unfriendly powers around the world.

David has a lot of stories to tell. He never told me any of them. Good spies keep their secrets. There was always a bit of the international man of mystery about Shedd. He grew up the child of missionary parents, spending many of his early years traveling throughout Latin America. In fact, on retiring, he said he planned to invest a good deal of time in missionary work. David is a man of deep faith. I know this work is very important to him, as important as his love of country and commitment to selfless service in his years of intelligence work. He is as good and honest a man as one can find in Washington, DC.

I don't know how he wound up being a spy. I never asked. I know he was an excellent one. I know that because anytime I ever meet anyone who worked in the intelligence community and I mention his name, they would give that little tight approving nod and say no more.

He has served as a visiting fellow at the think tank where I oversee foreign and security policy for several years. He was both a good friend and a valued colleague. He is an extraordinarily handy guy to have around. It is hard to do good public policy research on the intelligence community. Almost everything is classified. So it is great to have a veteran professional on call who can guide you in the right direction without giving away trade secrets.

Shedd also knows almost everyone worth knowing. That is crack cocaine to a networking junkie like me. I was in the military for twenty-five years before I started in public policy research, and although I had a clearance and knew a little about the business, I was like the guy who wins trivia night at the local bar compared to the reigning champion on *Jeopardy!* David Shedd is the Alex Trebek of the intelligence world. Being able to tap into his global web of family and friends was a real asset.

I mention David here because over the years through working with him, I engaged a lot more with folks working in the intelligence space and that led to an incident that became the genesis of the idea for this book.

Awakening

It was all due to happenstance. Across the river from DC proper sits Joint Base Anacostia. It used to be Bolling Air Force Base, which sat next to Naval Support Facility Anacostia, but in a rare inspiration of efficiency, the Pentagon merged them into one sprawling mess of buildings and antennas that wind along the far shore of the Anacostia River across from trendy new restaurants, expensive apartments, and the Washington Nationals baseball stadium.

One of the tenants at Joint Base Anacostia is the headquarters of the DIA, David's old hunting grounds. Through associates of David, I was asked to come over and speak to a group of intelligence officers from allied countries. As well as I could figure, this was part of an International Military Education and Training program. These are courses, schools, and seminars that the Defense Department sponsors for military personnel from other countries to assist in their professional development.

The DIA asked me to come speak on the "then" new hot topic—intelligence and social networking. Was I up for that? Maybe.

Sketching out what I knew about social networking online was easy (see chapter 1). I was not, however, an expert in intelligence tradecraft. Still, having written *Wiki at War*, I figured I knew a thing or two about the national security implications of social networking and could guess about how to fill in the gaps.

I never guessed how right I was.

Not far into our workshop, the conversation circled back to one of the important points that usually came up repeatedly in any dialogue I had on the topic (from my underattended pitch at South by Southwest to the present talk). This concerned the importance of understanding the linkage between the real and virtual worlds. They are, after all, not mirror images, where what happens in one is perfectly reflected in the other. Even though human dynamics dominate how both kinds of networks behave, the humans don't always behave the same way in both worlds. Just like on the *Sopranos* (and among real-world crime families), the Mafia in New Jersey and New York didn't operate in the same way, though they both followed the infamous Cosa Nostra "code of silence."

The virtual and real worlds often lash up imperfectly. An example that I often use is politics. Take the pop star Katy Perry. Before the election of 2016, Perry had about 76 million Twitter (now X) followers (today, she has over 100 million). She endorsed Hillary Clinton for president.[1] If all her followers had voted for Hillary and all the kids who pleaded with their parents to vote for Hillary had voted for Hillary, shouldn't Hillary have won? She didn't. Success in the virtual world doesn't guarantee influence in the real one.

Of course, celebrity endorsements don't always translate to political influence. That's the point. Assuming an action on any network has real-world impact is a big mistake. On the other hand, social networks can deliver. When Obama ran for president, he was the first to really exploit social media in a presidential race. It worked for him. Why did he get two terms and Hillary none?

Think back to the last chapter and the discussion of ISIS. The terrorist group ran a global recruiting campaign. To be fair, they did draw recruits from all over the world, but the overwhelming number came from just four countries—Tunisia, Russia, Turkey, and Jordan. Why is that? It looks to be a combination of reasons, many of them practical considerations, like the countries where ISIS had the logistical means to reach recruits and put them through a pipeline to get them to the caliphate in Iraq and Syria.[2] ISIS social networks had the most power and influence when they connected to extremist networks operating on the ground in the real world.

This was the point in the discussion when kismet came in, the real epiphany—or at least for me a total validation—happened. One of the officers in the course said, "I know exactly what you mean."

He told this story.

In his small central European country (which shall go unnamed here), it was publicly announced that US military convoys would be transiting their borders to conduct a military exercise in another central European country. Activity on the internet erupted. There were claims of violating the country's sovereignty. The nation would be drawn into a war with Russia. This outrage demanded mass protests and demonstrations. The Americans must be stopped. Roads had to be blocked. The people must rise.

The intelligence and law enforcement communities mobilized, not to secure the streets but to figure out what was going on.

They checked. Nobody was renting buses to take people to the protests. Nobody was reserving hotel rooms along the route of the convoy in unusual numbers. There were no public meetings to coordinate demonstrations. In fact, there were no signs of the real-world activities that would be expected if preparations for mass rallies were underway.

They studied the internet chatter. Sure, there were a lot of all kinds of people commenting back and forth. The instigators of the original call for protests, however, didn't come from these people. They came from bots.

Bots is short for robots. Not like Robby the Robot, the fictional icon from the science fiction classic *The Forbidden Planet* (1956). These were virtual robots, software programs or codes written to perform automated, repetitive, predefined tasks.

Let's talk about bots for a second. There are all kinds of bots on the internet doing all kinds of things, many of them perfectly legit. Traders, for example, use bots to buy and sell assets based on predetermined guidance. Chatbots interface with real-life users online, answering basic questions like "How is the weather?" or "What's your zip code?" They can be structured to respond in a manner that to users make it appear like they are communicating with another person. If you don't believe me, just ask your Echo. The plain fact is, bots, like Tom Hanks movies, are ubiquitous. Furthermore, virtually anyone can play in the bot world. You can download a bot app at the app store on your phone.

There is a dark side to bots. Bots can be used to infect other devices, using them as surrogates to run programs. These are called zombies. When zombies are dragooned on a large scale, taking over hundreds or thousands of computers, that creates what is called a botnet.

With a bot army at their call (either their own computers or the devices of innocents seconded to their command), bad people can direct bad bots to do all kinds of bad things. In social networks, they can mask as individuals conducting many nefarious missions. For example, bots have been employed in denial-of-service cyberattacks. Botnets can generate so much activity on a website or service provider that they overwhelm the system, denying other users access or just shutting the whole enterprise down.[3]

Bots can be used to drive up the traffic count on websites making them appear more active and popular than they are. Remember YouTube and how it pays influencers if their videos attract more users? Not surprisingly, bots are used to dramatically increase viewing numbers. This has created no end of issues for Google, which owns YouTube. Google, for instance, was charging for ads whether they were being watched by human eyeballs or virtual bots. That did not make advertisers happy. In turn, when Google cracked down on "invalid traffic"—in other words, fake views generated by bots—that impacted on what Google paid content providers. They got paid less. They weren't happy. To make matters worse, in 2018 security experts discovered botnets spread their influence by inserting malware into user devices that accessed YouTube videos, turning them into zombie computers.[4] That didn't make users happy. YouTube is a battlefield for botnet wars. If it wasn't for the fact that YouTube pulls in a ton of cash, Google wouldn't be happy either.

YouTube has been battling the bot problem for years. Their basic strategy is to identify and ban videos and YouTube channels whose numbers are driven up by fake views, because what they can't do is prevent the ever-ubiquitous bots from clicking on your charming cat video. Just do a search of "buy YouTube views," for instance; there are lots of sites that will pop up, offering thousands of view clicks for as little as a dollar. What a scam. From the "social" perspective, to users, advertisers, and content producers, these views are useless, as empty of value as Coke Zero is of calories. They don't generate comments or promote engagement with the material; they just ephemerally make something look popular.[5] As YouTube knows well, when bots are used as malicious tools of influence rather than internet service tools, they are poison to online social networks. When you take the human out of the network, as in the case of bots on YouTube, the social value is greatly diminished. Bottom line: Bad bots can be bad news.

Now back to the story. This is where the bots come in. Bots are much bigger than a YouTube problem. One of the biggest sources of botnets in the world is Russia. Russian cybercriminals control millions of computers that can generate billions of bots. Some of these botnets, we often suspect (sometimes we know), work with the complicity or at the direction of Russian military and intelligence services. I asked Shedd about this. He confirmed. He wasn't sharing any state secrets. There is a fair amount of open-source (unclassified publicly available information) literature about Russian activity. Even after retiring, working with private sector companies he saw the challenge of dealing with Russian malicious activity all the time, some of it directed from Moscow some of the time (a lot more on that later).[6]

In this case, guess what the intelligence service from our small unnamed central European country found? A forensic analysis revealed that all the internet rumors and fomenting for mass demonstrations came from Russian botnets. This activity had all the earmarks of a typical Russian disinformation campaign, throwing activity at communities online and just hoping something sticks, causing misery, confusion, and mayhem. In the face of this threat, the authorities took decisive action: They did absolutely nothing.

The US military convoys came and went. Hardly anyone noticed or cared.

Here is what these clever allies did right. They fused information from the virtual and real worlds to paint a coherent picture of influence and cause and effect. In the military, this is called gaining full situational awareness. A simpler way of saying this is, "Know your enemy." Like "taking the high ground," such simple military aphorisms have relevance. After all, armed conflict is competition, a clash of action and reaction between determined resolute foes. When social networking is enlisted in the battle for national security, or insecurity, it looks pretty much the same. The example that the allied intelligence officer offered up sounded to me like a masterclass in understanding the key dynamic in social networking warfare—focusing on whether the networks had the ability to link actions in the real and virtual worlds.

How can other networking warriors be as good as these guys? Here is the challenge for mastering warring networks. How can you structure networking activities to ensure decisive action on both digital battlefields and in the world of flesh and blood? There are lessons to be drawn from our ally's example. To be as good as my friends in an unnamed central European country, you must be able to (1) know what you should be fighting over, (2) know how to fight, and (3) organize that knowledge so that you can repeat those actions with predictable outcomes and achieve the results you want again and again. In short, spell out how to win. Here is how to do that.

What the rest of this chapter is going to do is to build a vocabulary for fighting networking warfare that can decisively impact online and offline, unpacking the answer to the question raised in the preceding paragraph. That requires wrapping a couple of strains of thinking together to create the concept of the networking way of warfare. These strands of thought include (1) how to conceptualize national security (know what to fight over), (2) how to draw lessons from other behaviors online (mostly social activism) for anticipating how to fight in national security competitions, and (3) how to adapt warfighting tactics (organize knowledge) from combat in the real world that would best work in the networking world. When these three sets of ideas are tied together, they lead to a list of eight activities comprising the fundamentals of warring with networks. Let's get started.

Get National Security Right

Online networks have influence when they connect to networks on the ground that are capable of operationalizing activities in the real world. If social networks can impact national security, we should take them very, very seriously. These networks ought to be the focus of effort for people who practice national security for one simple reason: That is where social networking crosses over into national security business. This is a crucial insight with important implications. It is worth digging into the implications. There are three.

(1) Of first importance is to make sure you are dealing with a national security problem. National security is a framework for dealing with competitions

between determined foes with opposing goals who are willing to use force or the threat of force to impose their will. There must be a bad person trying to do you in, and in turn, you must have the ways and means to influence their behavior, like kill or maim them, to win. These are "real" national security problems.

People are all the time wanting to define their pet issue (including overweight pets) as an issue of national security. Understand their thinking. After all, a national security issue is something that is important. It ought to take priority over everything else. It demands lots of resources and authority to address the peril. So, of course, advocates are anxious to make their concerns a matter of national security. But here is the problem. Because someone wants to treat a problem as a national security issue doesn't mean it is or it should be handled that way. It is like the adage, "When all you have is a hammer, every task looks like a nail." That, however, is no way to build a house.

Treating issues inappropriately as a national security challenge wrongly distracts efforts, encourages statist solutions, potentially undermines individual liberty, and in the end could exacerbate confrontation and conflict.[7] It is crucial to not screw this up. There is so much social media activity going on in the world every day, trillions of transactions, focusing on the right problem is so much easier said than done.

Let's take an example that dominates social media discourse: the case of climate change. Lots of people think climate change ought to be a national security issue.[8] After all, they argue, we could all die. Put the dying argument aside for now (there are lots of things that could kill us all from bird flu to planet-killing asteroids). Whether climate change puts people in danger or not doesn't necessarily make it a national security issue. National security is a structured way to deal with a particular problem set, appropriate for dealing with some difficulties but not others. You don't use a butcher knife to take out a splinter. You don't use national security to solve everything important. National security problems are about competitions. Climate change is about long-term weather patterns and their environmental consequences. You can't compete with weather.

Yes, I know climate change is impacted by human activity, but there are lots of ways to address human activity. When it comes to climate change, we don't fight with people over the climate (unless your plan is to wipe out large segments of the human population to reduce carbon emissions or make war over greenhouse gases). Merits of climate change policy aside, making climate a national security issue is an excuse to hijack national security infrastructure and the weight of national security's importance in support of an agenda, imposing solutions that are neither efficacious nor appropriate.

Meanwhile, the distraction diverts national security effort from its real mission—protecting the nation against enemies trying to undermine national security.[9] This can have serious adverse consequences. Indeed, treating climate like a national security problem might be creating problems for national security (I know that sounds a bit complicated; the issue is discussed in the epilogue). The point I'm trying to make here is simple: You don't practice surgery on the tennis court; they are two different things.

Looking at the interface between social and real-world action and their consequences help divining real national security issues from other stuff because connectivity can reveal the actions and outcomes of social media activity. With this knowledge, job one is then distinguishing what is appropriately a national security task.

(2) Put the priority effort first. Make sure first you address the issues that are most important. The precedence ought to go to trials that most directly affect your vital interests—the challenges that endanger the very survival of the nation. If more resources are available, you move on to important interests, ones that could potentially yield a significant advantage or repair a national weakness. Then move on to peripheral interests, challenges where a modicum of resources might be invested to achieve a modicum of benefits for the nation's security.

The taxonomy of interests (vital, important, peripheral) helps prioritize addressing the morass of national security mayhem that is often clamoring for attention. Here, looking at the connectivity between real-world and virtual networks also helps identify interests that may be at risk, opportunities, or problems worthy of attention.

Our convoy example discussed previously in this chapter is a case in point. Our allies in the small unnamed central European country had an interest in the transit of US military convoys. Why? Working with the United States showed they were good stewards of the North Atlantic Treaty Organization and reliable allies with Washington. Sending that message helped strengthen their own defensive posture. In addition, they were fed up with Russian interference in their politics and security. Not looking weak, dealing with Russian meddling was always an important matter. Furthermore, if the military convoys were disrupted, it would have been deeply embarrassing to the government. So, when the threat of public riots on the internet appeared, it was unquestionably something worth digging into as a national security issue. They put in the appropriate amount of effort and were rewarded with an appropriate, suitable result.

In short, looking at networks, what they do, and their capacity to intersect helps determine if there is a valid national security issue at stake and how much effort is worth investing in addressing concerns or opportunities. Cataloging interests correctly is a fundamental task to the national security business, like baseball players learning how to hit and field before they can play in the big leagues.

(3) In a national security competition, put most of your efforts where they matter most and have the most impact. Here is another military analogy. The US Army had an overwhelming military advantage during the Vietnam War. The Americans had more of everything, especially artillery. The enemy feared US firepower. When artillery shells started raining down from the sky, they could shred the ranks of an enemy column like a Cuisinart cuts through cabbage. As fearsome as US artillery was, you might remember that the United States lost the Vietnam War. Here is why.

The Vietnamese learned not to expose themselves to US military might if they could help it. For their part, the Americans didn't want to sit around in their firebases and do nothing but drink beer and bum smokes. No one wanted all that artillery to go to waste, waiting for the enemy to attack and die. So the Americans came up with a tactic. Artillery was used for what the troops called "H & I fires." That was army lingo for "harass and interdiction." The army fired artillery at where they "thought" the enemy might be. They blew up trails in the middle of the night, fired barrages at open areas where the Vietnamese might congregate for an assault on a base. Before dropping into a helicopter landing zone, artillery would flatten an area to take out any snipers. There was so much of this activity that US soldiers became oblivious to the constant crack and whine of artillery fire or the little shock waves of air that rippled their tents throughout the middle of the night. The drum of artillery was as constant as the sweltering heat and mosquitoes. Meanwhile, the Americans blindly lobbed rounds at map coordinates in the jungle, making fantastic explosions, killing a lot of palm trees and rice paddies, the occasional water buffalo, unlucky peasants going to visit their relatives in the next village, and occasionally, if often by sheer accident, perhaps an enemy soldier.

Artillery tactics proved as fruitless as a blindfolded man trying to solve a Rubik's Cube. When the enemy moved into an area, the first step they would take would be to do a reconnaissance and ascertain where the US artillery was and what kind of artillery it was so that they could determine its range. Then they drew a circle on a map out to the maximum range. Then, whenever possible, they made sure to conduct their maneuvers outside the circle. Unlike our allied intelligence officers from a small unnamed central European country who were smart enough to figure out how to not waste resources on a phantom threat, in Vietnam the US armies squandered resources big-time, never knowing if they were attacking a real enemy or not. What the army failed to do was take actions that delivered the maximum effect at the right time and place.[10]

Network competition works much the same. The biggest mistake to be made in social media is the belief that a flurry of activity, just like American artillery lighting up the night in the jungle, is producing dominating results. Looking at the capacity of real and virtual networks to interact and the results they produce helps target the attention of national security players on the places where they can be most effective.

So here is the bottom line: If you focus on real national security issues, if you prioritize efforts correctly, if you aim your efforts where they can have the most effect, and if you focus on the capacity of real and virtual networks to impact each other, your odds of prevailing in a competition go way up.

This is how you win in a socially networked world. To have a dominating influence in networking warfare, the nexus of real world and virtual world is the place to look and a guide for action. That is the key ingredient in the secret sauce of success.

Designing Winning

This gets us closer to delivering a real catalog of winning ways for becoming a dominating network warrior, but we are not there yet. We need the right tactics.

Tactics are not a list of what to do. Tactics are a menu of how to do things—practical, concrete, realistic, and executable instructions. Tactics are also more than aphorisms.

It is easy, for example, to say, "Take the high ground." The Confederate army tried that at Gettysburg. They lost. The Jews held the high ground at the battle of Masada (73 BCE). They all died. The GIs took the high ground in Normandy at Omaha Beach (1944). They won.

It is easy to say, "Know your enemy." The United States spends more money on intelligence than any other country in the world, and we get surprised all the time. Sometimes we win anyway, like the Cuban Missile Crisis (1962). Sometimes, like the Bay of Pigs (1961), we get our butt kicked.

It is easy to say, "Be strong at the decisive time and place." We tried to do that in Iraq and Afghanistan. We bested Saddam Hussein in 1991 and defeated Iraqi insurgents in 2011. We tried that against the Taliban. In the end, we lost.

Glib maxims are useful, but they are like chapter headings in book, just an introduction. They are not enough for real fighting, in the real world or on digital battlefields. For tactics to be usable by networkers, they must be a clear and effective guide to actions that offer ways for networking warriors to do the right things at the right time to win. Good tactics are not built out of thin air; they are based on practical experience. So the first challenge is finding lots of practical experience. Here is what I found.

To unpack a tactical scheme that builds dominating social networks or makes your enemies impotent requires going back to how networks function and how and why they intersect in the virtual and everyday worlds. There is a ton of research on this. Let's see what the last decade has taught us since *Wiki at War* hit the bookshelves (and audiobooks and Kindle). One of the best sources is case studies that explore the psychology of online activism and social movements.[11] The reason why looking at this research is particularly fruitful is because (a) there is a lot of data to study, (b) much of the activity crosses over between the real and virtual worlds, and (c) this is an area where researchers have invested a lot of effort studying the interactive relationship between the online and offline worlds.

Unpacking Social Activism

Slacktivism, an issue mentioned in chapter 1, is one of the frustrating challenges in harnessing online social activity. The last chapter described the case of "Save Darfur," a social media campaign that did nothing to save Darfur (refugees suffering from the ravages of civil war in Sudan). That is a clear example of where the online and offline activity did not materially impact each other. This is called slacktivism. The term "slacktivism" comes from putting together the words "slacker" and "activism." The moniker is intended to describe online activities supporting political or social causes that are characterized by evoking very little effort or commitment. This is the offline equivalent of walking by a homeless person, thinking "I feel your pain," and then just keep walking.

Social activist free riders, like slacktivists, are only one of the issues for understanding how behaviors in the real and virtual worlds link (or don't link) together. Researchers have looked at all kinds of perplexing questions. Is online activism promoting just feel-good "liking" on social media, or does browsing generate real social activity including "documenting and collating individual experiences, community building, norm formation, and development of shared realities"? Does online activity trigger collective action in the real world? Can social media hinder rather than help action and protest? One extensive review of all the literature and all these weighty questions concluded that the record is, well, mixed.

In trying to unpack how the two worlds connect, the authors described three kinds of relationships.

(1) Classic slacktivism where the online world did not appear to influence offline collective action. The incident of social media protests and the US convoys was a good example. While many people followed the Russian bots adding their anti-American rants, none of them charged into the streets to throw themselves in front of US tanks.

(2) Digital dualism where online and offline protests are both present but not connected. This was often the case in the Arab Spring, with events on the street and vigorous posting and debating among the Arab diaspora going on at the same time, in parallel with one another but not impacting one another.

(3) Active action where virtual and real-world activities positively correlate. A clear example of this was our Afghanistan crisis action group. The members of the group heavily leveraged their virtual and real-world activities in pursuit of a common agenda: helping people escape the Taliban.

Of the three relationships, the authors concluded, "Most empirical evidence suggests that online and offline activism are positively related and intertwined" but—and this is an important caveat—not always.

There is a reason that positive correlations of online and offline activist networks often occur. That has to do with a point that was raised again and again in chapter 1. Networks, where the activities are driven by human actions, whether they are in the real world or in a virtual space, act like human networks. Humans instinctively form networks to get things done through collective action. In examples from pre-nineteenth-century peasant food riots to contemporary social media campaigns like Chick-fil-A's the Little Things (2018), "calls to action cascade through interconnected personal networks." That this holds true for activist networks just makes sense. We would expect these virtual and real-world networks to affect each other.

There is no question that networks can spark action. Here, however, is the point where things get murky. There is no consensus in the research about what dynamics trigger the proliferation of activity between networks operating in the physical world and in virtual spaces. Why do some converge and others do not?

The reason why the interface between online and offline networks is not always the same is because humans don't act consistently either online or offline. For instance, the researchers cited studies where digital dualism dominated. Some users adopted different personas on the internet from how they acted in the real world. "Relatively anonymous online environments free people from concerns to be positively evaluated and consequent social restrictions to their behavior," the research they reviewed suggests, facilitating, "activism [online] without fear of social repercussions." The fact that anonymity can affect behavior is not surprising. In the example of social activism China (described in chapter 1), activists highly constrain their behavior because the government does not allow anonymous social activity and activists know they might be held accountable for opinions the government doesn't like.

What the research suggests is that virtual behavior bleeds over into the real world, breaking down digital dualism, when the norms of a group that interact online behave in the same way in the real world. For example, users who are anonymous or who use a pseudonym to engage in an online community "are likely to riot if that community consists of violent activists but disorderly behaviour is less likely if their community consists of pacifists." In other words, whether real and virtual worlds converge or diverge is "context dependent," depending on the capabilities of the networks and all the factors that influence their reach and effectiveness.

Take the example of two Chechen immigrant brothers, Tamerlan and Dzhokar Tsarnaev. From outward appearances, daily these young men were becoming more everyday Americans. Dzhokar captained his high school wrestling team. Tamerlan dreamed of wearing red, white, and blue shorts, boxing in the Olympics. Both attended college. On April 15, 2013, two bombs improvised from pressure cookers, nails, and explosives planted by the brothers detonated along the route of the Boston Marathon.

The terrorist attack killed three innocents and injured scores more. The radicalization that motivated the attack came from engaging with groups that did not just profess a radical Islamist ideology, but they also practiced it and their commitment to violence in real life. One of the key influences was the sermons of radical preacher Anwar Nasser al-Awlaki (that the brothers accessed online), who was a key operative of al-Qaeda helping mount the 9/11 attacks and who later became a propagandist and recruiter for ISIS.[12] The Boston bombers case offers a powerful example of how the direct connectivity between like-minded virtual and real-world activities works.

Unfortunately, the attack on the Boston Marathon doesn't offer a prescriptive guide for understanding how the dynamics between online and offline networks play out. The Tsarnaev case is forensic. In other words, the impact was only understood after the fact, after people died, and chaos reigned in a US city. For tactics that focus on the linkage between the real and virtual worlds to be helpful in dominating social networking, there must be a way to assess the strength of connectivity in time to impact network performance.

For years, folks have been squirreling around looking for tools to predict which social networks were going to trigger actions from massive fund-raising to high school mass shootings. For example, a study of mass public protests in Ukraine used real-time geotagging of online activity to predict whether virtual and real-world actions were converging and putting people in the streets to protest.[13] Research suggests that given the right data and circumstances, it is possible to correlate virtual and real-world activity and predict outcomes. The challenge in duplicating the results of this research is finding relevant activities to measure and correlate in real time. That can be a real challenge.

Another issue is determining what counts for influence. Social media that brings howling rioters to the streets or triggers two brothers to build bombs and kill innocents to satisfy their outrage offers obvious evidence of impact. Other activities, however, could be more subtle, delayed, and defused and hence more difficult to measure. In his research, Aaron Noland, an expert in communications studies at James Madison University, concludes that social networking can manifest in a myriad of social-cause engagement activities. He even questioned if research could demonstrate that there are true "slack-tivists" who are stimulated in no concrete way whatsoever by their online engagement.[14] Can you prove a negative? Can you really prove that in response to social media, no one does anything ever? One man's slacktivist might just be another man's slow learner. Measuring impact can be tricky.

Take the example of the Ice Bucket Challenge (described in chapter 1) where individuals were encouraged to video themselves being doused with buckets of ice water to help raise money for charity. At the time, the social media campaign was dismissed as a slacktivism stunt, with many people engaging in the activity but without proportional real-world benefit. Most people participated, critics concluded, because it felt good, not because they were motivated or committed to do good. Subsequent assessments, however, found that the campaign had a significant and lasting impact on social mobilization as well as being linked to increasing donations and grants over the long term.[15]

Here is what this all adds up to. The common thread in what the mixed salad of research says about social media and how it connects to the real world is the human factor. The strongest linkages are going to occur where the networks most strongly reflect the elements of strong human interaction. Networks that amplify human behavior attract each other like iron and a magnet. The more pronounced the human aspects of the network and the more effective they are, the more powerfully they draw together across the divide of the planes of virtual connectivity and reality.

To summarize, this is what we know from looking at social activism as an example of understanding the linkage between the real and virtual places where humans hang out. (1) If we are looking to devise tactics to dominate social networks and how they influence life in the real world, we are looking in the right place. Social activism is a great exemplar of how humans connect online and offline. (2) Establishing a link between the real and the virtual world doesn't necessarily ensure impact. (3) Context matters. Different factors can affect how well links secure a strong crossover between online and offline activities. (4) It can be difficult to proactively identify and measure the impact of social media on real-world activity. (5) The way to build virtual and real-world networks that bind is to focus on maximizing influence on human behavior in the networks. The capacity of the networks to generate real action is a key variable. All this adds to an intellectual tool kit for building a set of actionable networking tactics. This approach is way better than just handing out bumper stickers like "Take the high ground."

Technology, Games, and Tactics

Unpacking social activism helps lay out all the issues that must be addressed in formulating usable, dependable tactics for social networking warriors to war with. What is particularly interesting is that these dynamics transcend individual technologies. They virtually all relate to human behavior. This reinforces a crucial requirement for social networking tactics: Focus on people, not technology. Social networking is fundamentally not a contest of technologies. It is not "our electrons" versus "their electrons." Tactics ought to focus on the behavior of the humans. Impacting their choices and behaviors is the goal of social networking warfare. That impact requires acting in the real world, not just online. Bottom line: You need tactics that will impact actions on the ground.

In a competition, devising good tactics always starts with asking how to best influence the competitors. Developing artillery tactics in Vietnam ought to have started with asking the question like "How do I defeat the enemy?" not "What can we do with artillery?" Dominating social networking raises the same kind of issues. Asking the right kind of question leads to the right kind of answers. The question is not "What do you do with technology?" The question is "How do you affect the behavior of networks?"

Trashing technological determinism is not to claim that technologies cannot be a tool of tactics. Of course they can. Tactics, for example, changed when armies confronted machine guns and artillery barrages in the outbreak of World War I. Otherwise, every soldier would have been dead before Christmas. Still, before looking at the technology, first focus on the players. This is particularly important in national security competitions. National security is like chess (in part) in that it is a test between competitors. Networking warfare is in another way, however, not a game. There is a crucial difference. Like technologies have physical limits, games like chess have procedural limits. Chess, like all games, are bound by rules. In contrast, one of the dynamic aspects of national security competitions is that the players make the rules. They can honor them, bend them, break them. In a game of chess, one player can win by picking a cleverer stratagem. They can play the Queen's Gambit or Vladimirov's Thunderbolt, but they are still following the rules. If chess were national security, one player could pull out a gun and shoot the other in the head. They win. That is the name of the game. Humans are only limited by their guile, wits, conscience, and perseverance.

It is better to formulate tactics that don't begin with looking at the artificial things that bound competition like "rules" and technology. Start with the human factors that influence competition, the raw competitive instincts—how humans clash and cooperate. The other stuff can be folded in later.

Take, for example, phishing or deepfakes, malicious cyber tactics and technologies that we have discussed before. They work because they exploit flaws in human decision-making. The technologies didn't make us do stupid things online. Social engineering succeeds above all because competitors know humans sometimes do stupid things and they went out and found or built technologies to exploit that.

An action-oriented way of formulating tactics also explains why tactics are guides to action, not formulas to follow. If you follow a recipe to the letter, you will produce a soufflé every bit as good as Anthony Bourdain ever could have pulled from an oven. If you follow tactics to the letter in a national security competition, you still might get your throat cut. The simple reason for that is the other side might have developed better tactics (or they just got lucky and had a sharper knife).

Let's look at an example. In 2009, the world was stunned by the Moldovan "Twitter revolution." The tiny state at the edge of Europe for years was firmly under the thumb of Russian influence and corruption. The post-Soviet era didn't seem very post in Moldova. Seemingly out of nowhere, protests emerged online and offline, delivering unprecedented political reform.[16] Then, in 2021, reformists led by Moldovan President Maia Sandu swept parliamentary elections, securing a shocking political mandate to combat Russian influence and public corruption. Wait a second, didn't the Twitter revolution already do that? The answer is yes, but after 2009, neither Russia nor the corrupt oligarchs threw up their hands and conceded the game. They fought back with their own tactics to undermine Moldovan democracy, security, and independence. President Sandu's political victory in 2021 was another dramatic reversal for Moscow but almost immediately matched by another Russian counterpunch. In the fall of 2021, Moscow suddenly decided to impose a dramatic increase in gas prices and restricted gas flows to the country. The Russians then demanded political concessions to turn the gas on and drop the price. The fight is far from over. There is a lesson here: Tactics are just a tool to win. They don't guarantee a win. Just ask the Moldovans.

Still, like never getting a job you don't apply for, you can't win without fighting. You can't fight well and smart without good tactics. *Digital Dominance* is going to teach you how to fight for the top with a set of tactics built around how to influence human behavior.

Designing the Handbook

Let's add some more ingredients to the secret sauce of dominating social networks. This is about way more than competing for higher rankings on Reddit. A lot of the advice on maximizing social media, recommendations that focus on the superficial aspects of designing a web page, is useless for what we are trying to do. Dominating networks is a lot more than getting clicks and open rates up. We can, and must do, way better. *Digital Dominance* is going to explain to you how to bridge real-world and virtual networks and make things happen.

Not to overdo military analogies (but then, why not? This is, after all, a book about national security competition), I think looking at network warfare as a tactical problem is useful. Part of that is my background. Before getting a cushy think tank office, I had a real job. I served in the army for twenty-five years. It shaped how I think about things. It's not just me. David Shedd talks about challenges in the same way. That is because the military and intelligence community share a common perspective. We look at problems as competitions. We assess what we must fight with. We consider the motives, intent, and capability of our adversaries. We consider the terrain we will be fighting over. We focus on how to win. In the army, I trained to win battles on the ground. In his career, Shedd battled to make our spies more effective and their spies less so. Dominating networks is a fight to win in digital space and physical space. In the world of competition, good tactics work well in all these places.

There are a lot of suggestive analogies in the real world for national security social networking warriors. One of the best examples is the battle against the use of improvised explosive devices (IEDs) by insurgents after the Iraq War. If you are not familiar with this threat, go watch the Oscar-winning movie *Hurt Locker* (2008). To meet this challenge, the Pentagon set up an organization called the Joint Improvised Explosive Device Defeat Organization (JIEDDO). JIEDDO recognized that the "threat" was not the bomb but the network that made the bomb, a threat that compromised a massive system including the people that funded operations, the ones who directed operations, provided supplies, assembled the bombs, scouted the targets, planted the bombs, videoed the attack, and paid the bombers. The insurgents did everything they could to protect their network. JIEDDO's job was to come up with tactics to counter them. This was network warfare and it's a model for fighting socially networked war (more on JIEDDO later).[17]

So let's prepare to be good fighters for winning on digital terrain. Let's spell out the elements of what makes for good tactics. Taking what we have learned from the dynamics of online social activism and based on our experience of what works in the military and intelligence space, adapt the most applicable and useful concepts that would help in mastering social media competition.

For starters, begin with the point that was hammered on earlier in the chapter: Focus on what is most important. When it comes to planning a campaign against an enemy, this is called attacking the "center of gravity." This term of art was popularized by Carl von Clausewitz, the famous nineteenth-century military philosopher and author of the *Art of War* (first published in English in 1910). Clausewitz used the term (*Schwerpunkt* in German) about forty times (sadly with no consistent meaning). Still, it must have been important; otherwise, he would not have kept harping on it. In a nutshell, Clausewitz in various ways described a focal point such that if a campaign made the object of effort, then operations could facilitate empowering one side or destabilizing the other.[18]

For example, during World War II, the army air forces argued that German ball-bearing plants in Schweinfurt were crucial to the war effort. If the factories were wiped out, Germans couldn't continue to manufacture airplanes to make war. Without air cover, the Nazis would be defenseless against Allied bombers. The bomber boys could then pound Hitler's Germany to dust without much interference. The war would be won; ticker tape parades and dancing to follow. At least that was what the bomber boys thought (bomber boys always think they can bomb their way out of any problem). They went after the German plants with a vengeance, Operation Pointblank (1944). In competition, if you want to win decisively, go after the center of gravity.[19] So, tactics should help identify and focus on the most important targets.

The next point gets to the point of how much effort, blood, and treasure do you expend in going after what's important. The answer is, enough but not too much. Let's return to Clausewitz.

In March 1812, a thirty-two-year-old Prussian officer, Carl von Clausewitz, crossed the frontier into Russia, intent on joining the imperial army and continuing the fight against an old enemy. In 1806, Napoleon (Bill Gates, the founder of Microsoft, by the way, loved to read about Napoleon in high school. Just saying) humiliated the Prussians at the battle of Jena-Auerstedt. Clausewitz suffered a further ignominy having to accompany his defeated general to France as an aide-de-camp. Fresh from the indignity of being a prisoner of war, Carl was itching for payback. Napoleon stormed across Russia and marched into Moscow, just as he had in Berlin. But the army of the Romanov emperor, while beaten, was still intact. And the emperor refused to surrender. With his starving troops facing a punishing winter, Napoleon began the long, painful retreat to France. His army dissolved. He lost his crown. Clausewitz saw this happen firsthand (fighting for the Russians). That experience was the seed that sprouted another important idea.

In unpacking the history of the Russian campaign, Clausewitz introduced the militaries of the Western world to one of his most enduring and powerful insights, what he called the "culminating points of victory." Clausewitz had originally described the concept in a separate essay but then added and supplemented the material in book 7 of *On War*. Clausewitz thought the culminating point offered a profoundly important discernment into how wars are fought. Even when one side had every advantage over the other, victory could not be guaranteed. "Often," Clausewitz cautioned, "victory has a culminating point." Taking the battle to the enemy required expending resources—men, ammunition, fodder. The more an army advances deeper into enemy territory, the more it moves away from its base of supply and support, potentially diminishing its combat power. "Every reduction," Clausewitz warned, "in strength on one side can be considered an increase to the other." At some point, a powerful attacker might expend so much in pursuit of an elusive objective that their force becomes weak and vulnerable to counterattack.[20] When the force crosses over the "point" where the advantages that might ensure victory are whittled away, the force crosses the culminating point. That is exactly what happened to Napoleon. He stretched his army across Europe like a rubber band, and then straining, the band snapped. The Russians chased him all the way back to France. The lesson learned is, don't put so much effort into winning that you make yourself vulnerable to losing.

So that raises another question: How do you influence the most important factor in a competition and not risk defeating yourself in the process? The answer is, no surprise, it depends. There are usually two ways to solve any military problem—directly or indirectly. Direct is just going straight at it. The Union army at Gettysburg held the high ground. The Confederates tried taking it. That was a direct approach. Indirect is to not go directly at your target, purposefully challenging the competitor's strength. An indirect approach at Gettysburg would have been to outmaneuver the Union army, forcing those "damn blue bellies" to abandon their commanding defensive position. Neither approach is inherently right or wrong—it depends on the context surrounding the choice. That is what good tactics are about. Good tactics offer different approaches to solve a problem, making it harder for your adversary to predict exactly how you are going to come at them.

More military stuff. There are also two ways to fight. You can go after the other side; that is called offense. Or you can protect yourself against attacks from the other side; that is called defense. At Gettysburg, the Confederates were on offense—attacking the Union positions on the ridge behind the town of Gettysburg. The Union was on defense, blasting away at the attacking columns of rebels.

Neither form of fighting is always the right answer. The Confederates were on offense at Gettysburg and lost. That same summer, they were on defense at Vicksburg and lost. Not picking on the Confederates (even though they were on the wrong side of right). The Union lost plenty of battles—both attacking and defending. Like with most tactics, it is about picking the right one at the right time. The right answer is the right answer. The bigger point here is, tactics should cover offense and defense. In competitions, most competitors wind up doing a bit of both.

And here is, I promise, the last military comparison (at least for now). This point goes back to Clausewitz once again, who wrote, "Everything in war is simple, but in war the simple is difficult." Lots of forces and factors can mess up what happens on the battlefield. When the Americans hit the sands at Omaha Beach in Normandy, everything went wrong, but they got some simple things right and still managed to take the high ground. The lesson is that the real world is messy enough; simplicity in design and execution increases the odds of bumbling through all the obstacles. Tactics should be simple. Execution will be difficult enough.

Let's sum this up. Tactics are a playbook, not a textbook. They should (1) impact the critical, (2) don't overdo it, (3) address direct and indirect options, (4) cover offense and defense, and (5) keep it simple. It's just that simple.

If we put together what we know from military confrontations that make for sound national security practices and we add what we have uncovered about how social media and networks work, we have all the conceptual tools we need to build a networking warrior playbook.

What follows is a set of actions that can incorporate all the elements of tactics that were outlined above. Below is a summary. The chapters that follow go into a lot more detail on how to conduct offensive and defensive actions and all the other elements of tactics that can be used in networking warfare. For now, here is the *CliffsNotes* version to get things started.

Dominating

It is time to circle back to where we started. Let's start with why our little social network (discussed in the prologue) had the capacity to influence events in the real world in the tumultuous last days of withdrawing from Afghanistan. In retrospect, it was because our gang had all the components of a network that can operationalize activity in the real world, the components that build capacity by touching directly on the key human aspects of networking. I am going to list them here. In subsequent chapters, *Digital Dominance* will unpack each element. In the following chapters, we will build out each of these ideas describing the tactics that are most helpful (based on what we laid out in this chapter) for guiding how to confront an opponent and win.

Digital Dominance outlines eight actions for dominating social networking, starting with four actions for building, sustaining, and fighting with an operational network.

> (1) Recruit the Network. Networks won't work if the humans you pull into the network don't have the right skills, knowledge, and attributes—in other words, the ways and means to do the right stuff in the real world. It might not be essential that they have this collection of capabilities when they enter the network, but they must have them when it is time for the network to go to work. What is important is that it is not how big, diverse, rich, talented, or influential the network members are; the issue is, are they the right members for what the network needs to do? Whether the goal is to build a large network that operates in broadcast mode or a small network for collaboration in conversation mode, the task is still the same. Get the right people in the network. Let's go back to Katy Perry's army of 76 million Twitter (now X) followers. How many of them had the full set of tools and the commitment to help win a presidential campaign? The answer is, not

enough, or Hillary Clinton would have won and probably still be president. To dominate a network, you need the right networkers.

(2) Build a Value Proposition. What keeps people coming back to a social network is the satisfaction that they are getting something out of the network they want or need. Let's go back to the example of the Tsarnaev brothers. Frustrated, isolated, and angry, through a web of radical connections they found the support and motivation required to become real terrorists. To make a network work, give people a reason to want to come back and engage, to be part of the club.

(3) Operationalize the Network. There is no use building a network if you don't do something important and impactful with it. Let's go back to the Ice Bucket Challenge, which was initially dismissed as just a slacktivist playground. The ALS Association used the attention raised by the virtual activity to convince funders that they needed to triple the organization's annual funding for research.[21] Dominating networks do things.

(4) Build a Feedback Loop. Productive networks have the capacity to measure steps 1–3 and use that input to build more capacity, capability, and resilience into their network or at the very least determine if they are achieving the effects intended. Going back to our Afghanistan crisis response group offers an example. At one point, I just emailed everyone and asked if the network was still useful. That was a simple means of verifying that the network was delivering operational impact. Other sources of feedback included monitoring the email traffic—reading about people connecting and coordinating and reporting back to the group about the impact they were achieving. The more you use feedback to strengthen the network, the more it dominates.

Together, these four deliberate steps describe a simple set of ways for building dominating networks, the kinds of networks that deliver real-world impact.

Building strong networks, however, is only half the battle. Remember, networking warfare is a competition. Dominators will also seek ways to diminish the capacity of their adversaries to be social networking warriors, while recognizing their adversaries could well be using the same tactics to go after them, which means also developing tactics to protect your networks.

There is a vast literature on how to take down networks, particularly in the national security space, much of it directly relatable to social media competition. David Shedd and other friends and colleagues, by the way, were a big help in pointing me in the right direction. Beating your enemy can be broken down into four steps.

(5) Map the Network. Before going on the offense, it makes sense (unlike General George Custer) to know what you are dealing with. Who is participating? What are they doing? How are they doing it? What impact are they having? This step is a prerequisite for identifying capabilities, strengths, and vulnerabilities and then devising the best tactic to go after the network. In the case of JIEDDO, for example, once they scoped out the chain of an IED threat, they focused on getting to the "left of the bang."[22] In other words, the further up the chain they interdicted activities, the more IED attacks they prevented. Best case: They stopped the people funding the whole enterprise. Worst case: They tried to devise means to mitigate the force of the explosion. The bottom line is, you need a map before you can know how to decide to get to where you want to go.

(6) Disrupt the Network. One way to prevent an adversarial network from dominating is to interfere with their activities—thwarting recruiting, contorting interactions, or unsettling activities. For instance, as online radicalization activities like those that recruited foreign fighters for ISIS or inspired the Tsarnaev brothers increased, major social media companies invested effort in deplatforming their content, making it harder for these people to connect with one another. Make the enemy have a frustrating bad networking day, losing trust and confidence in their network.

(7) Defeat the Network. If a network cannot accomplish its mission, it is not a network worth worrying about. Even though the US forces in Iraq never managed to stop every IED attack, they did muster up a plan that defeated the insurgency employing the IEDs. That took care of that problem.

(8) Destroy the Network. In the end, the best way to take a network down may be to take out the network, its members, or the supporting infrastructure. If the network is crushed, there is no networking. In 2011, the US government directed a drone strike that killed Anwar al-Awlaki, ending his role in the ISIS social media campaign.

Based on the assessment of what would work best against your competitor, these steps might be taken individually, in combination, or in sequence. Like all tactics, it is a matter of judgment and choice.

Also, don't forget that the tactics used to eliminate competition can also be used to help guide countermeasures—in other words, determining ways to prevent your adversary from taking these steps against you. This is another reminder that social networking warfare, like real war, is a contest of action and counteraction.

Off to War

Chapter 1 summarized *Wiki at War* and what's been learned since. This chapter laid out where we go from here. It explained what made for impactful national security social networking. These recommendations come from understanding what national security competition is and fusing those ideas with the practices that stimulate networks (looking at social activism as a model). Then add to this the tactical concepts that would best facilitate impact in the security space.

The chapter also described what tactics are needed to be the dominant player. Based on all that analysis, *Digital Dominance* detailed a set of eight actions that ought to comprise the playbook of the social networking warrior. I might never have got to this point if I didn't share stale bread across a checkered tablecloth from David Shedd—the serendipity of just lunch.

Long after he retired, by the way, Shedd came out of the shadows and wrote an influential book about Chinese malicious activity stealing intellectual property. He is still in the business of going after adversarial networks, I guess.

So what's next?

Now, having assembled the tools for networking warfare, it is time to dig in and explore how to use them, how to put tactics into practice. The next part of *Digital Dominance* allocates a chapter to each of the actions, covering how-to. Think of it this way. If this chapter explains how to shoot, the next chapters go to the digital network firing range for target practice.

PART TWO

How to Build Networks

3

DRAWING A CROWD

Jim is a talent—no pun intended.

A good friend, colleague, and mentor, Jim Talent, maybe more than anybody, taught me the importance of building impactful networks. The fact is, nobody knows how to draw the right kind of crowd better than Jim.

After serving a term as a senator from Missouri (2002–7), Jim couldn't let go of his passions for policy and politics. So, while he started a real job, consulting in the private sector, he also maintained affiliations with Washington think tanks to continue championing the causes that he believed keep the United States free, safe, and prosperous. He was a hit in the think tank world. Senator Talent brought the whole package, the political skills that only get honed by plowing past sharp elbows in Washington hallways, unmatched expertise in national security issues (far more than the average politician), and the practical how-to knowledge that works in real workplaces.

The descendent of Russian-Jewish immigrants who settled in Missouri, Talent in his presence and demeanor is all "show me" state. Tall, dark-haired, and handsome, he was born looking like a senator. After starting a law career, Jim served four terms in the Missouri House of Representatives. In 1992, Jim ran for Congress and won, even though he was facing an incumbent who heavily outspent his shoestring campaign. In 2000, Talent ran for governor. In 2002, he ran and won a seat in the US Senate.

As a freshman, Talent got a coveted assignment on the Senate Armed Services Committee. There, Jim demonstrated a passion for a select number of vital issues of the day. One of them was national security. He remained plugged in on defense and security matters in and out of office. Talent was vice chair of the bipartisan Commission on the Prevention of WMD

Proliferation and Terrorism. He also served on congressionally chartered independent panels that reviewed the Quadrennial Defense Review of the Department of Defense (2010 and 2014) and is a member of the congressionally chartered US-China Economic and Security Review Commission.

We started working together when Jim entered the think tank world after the Senate, sharing a common concern: the state of national defense. At the time, the United States was six years into the War on Terrorism that erupted after September 11, 2001. While the country had spent billions on defense, virtually all those dollars went to paying for operations in Iraq and Afghanistan (everything from ammunition and payrolls to fixing tanks and airplanes). The US government had not invested in modernizing the military in over twenty years. If the Pentagon were your car, this would be like going decades buying gas, paying tolls, and nothing else.

Before Ronald Reagan's presidency, the military went through a similar experience. During the Vietnam War, the United States took a "peace dividend" and didn't invest in refurbishing the military. As a result, the armed forces went "hollow," lacking training, readiness, and updating of equipment. That dismal state persisted until Reagan fixed the problem with a jolt of increased investment in the Pentagon. Reagan called this approach "peace through strength." A group of us, including Talent, believed we needed a similar investment after the long war fighting al-Qaeda and their ilk.[1]

When a new president was elected in 2008, the administration, rather than investing in rebuilding the armed forces (as we had hoped), opted for another peace dividend. Predictably, military readiness and capabilities soon began to decline. We were heading hollow again. Our goal was to turn that around.

The think tank where I work is a nonpartisan, not-for-profit institution prohibited from engaging in political activities like elections. What we were not prohibited from was building a network of like-minded national security professionals committed to fleshing out how to prevent the military from becoming hollow again. If someone who cared about national defense won the presidency, we also were not prohibited from advising them on how to fix the mess the military was in.

Jim was our guiding light for building a volunteer network of hundreds of dedicated professionals. All that brainpower turned out to be useful for something called the transition team.

Reins of Power

Presidential transition teams are a modern political creation. Unlike parliamentary democracies where the parties out of power have "shadow ministers" who monitor government functions (like defense) and are prepared to step in as ministers if their party is ever elected to power, in the US presidential aspirants don't pick their cabinet until after they are elected. Furthermore, in parliamentary systems the ruling party usually just enters government with a handful of party officials. In the United States when a new president enters office, they get to make about four thousand political appointments. In the US system, the transition of leadership from one party to another is just a much more complex deal.

Before 9/11, presidential candidates kept their preparations for taking over the White House low key. There were several reasons for that. No candidate wants to distract time and resources from the focus on campaigning to win the White House. Also, as soon as candidates start talking about the officials they'll appoint and what they will do, that becomes news and distracts from the candidate and their campaign. Finally, if candidates start filling out their cabinet and policies before the election, they could well start alienating and losing the enthusiasm of key supporters who had hoped that maybe "they" would be the influential ones in the next administration.

After 9/11, Congress started to think that perhaps presidential transition might be a real vulnerability. What would happen if a terrorist incident or crisis occurred right after a president came into office—just putting their team and policies together, still trying to get people security clearances, and only beginning the appointment of high-level officials that had to be confirmed by the US Senate? So Congress started thinking about a law that would establish federal support for transition activities that would begin after candidates were nominated by their parties at their national conventions and continue through inauguration day. These transition activities would be considered nonpartisan, fire-walled from campaigns. By having these teams, Congress reasoned, a new president would be better prepared to lead on day one.

The Pre-Election Presidential Act of 2010 provided authority for significant resources for the candidates of both parties to establish transition teams. The 2012 election was the first time that those authorities were fully employed.[2]

Our goal was to make the most of the transition, an opportunity to turn defense policy around, stopping the hollowing of the military before it got worse. Through our network, we assembled a small army of people and policies for the transition team. This networking bridged the virtual and real worlds, with groups operating online and offline. A number of these individuals made their way into the official transition team. Jim Talent was a key national security adviser and could well have wound up as the secretary of defense if the other presidential candidate had not won reelection.

Jim taught me a valuable lesson in the 2012 campaign. Based on that experience, our think tank helped build an even bigger network for the 2016 election cycle. Our small army became a bigger army. Thirty of the folks associated with our think tank served on the official presidential transition team, including me.[3] There was also no question that the network was having an impact, particularly on defense. When the candidate made their major speech on defense policy, it drew heavily from our research.[4] (The person tasked with drafting the speech also happened to be part of our network. I know the speechwriter pulled from our stuff because he told me so.) In the end, we got an administration that embraced peace-through-strength policies we believed were so important.

Despite our best efforts, there was still lots of chaos and controversy in the 2016 presidential transition. It turned out, for instance, the individual selected for secretary of defense had many disagreements with the president (he even kept his transition team separate from the president elect's). He didn't last long. Still, a lot of the policies the president insisted on matched well with what we were looking for.[5] It may have been messy, but the defense policies in the end were sound.

The chaos of transition and the tumultuous times after inauguration were just another reminder that the arena of policymaking and personnel can be just as chaotic as real battlefields. A famous German general once said, "No plan survives contact with the enemy." On the other hand, Dwight Eisenhower (a US general who beat the Germans in World War II) said, "Plans are worthless, but planning is everything." A plan, a lifeline of a guiding idea, helps navigate the bedlam of real-life competition. I can't help but think our planning helped. Our plan to have a network paid off (more on planning in chapter 5). People who served the newly elected president told us so, more than once.

In retrospect, the reason this effort succeeded from the start was because we recruited the right crowd to participate. Much credit for our success goes to Jim Talent. In 2012, his street "cred" on national security was well established and that credibility attracted a lot of folks who felt if Jim was involved that this was a serious effort worth supporting. In 2016, we were a known quantity based on our 2012 efforts. Furthermore, we had made a clear commitment. Our network wasn't there to pick winners and losers. The network was there to serve any leader interested in people and policies committed to a strong national defense. This was a club worth joining.

I raise the transition support network here not to claim credit for anything or suggest this group played some outsized role in shaping policy but to highlight what I learned from the experience. Through four presidential cycles (we also had a network up and running in 2020, and our foundation established a network of over eighty organizations to support transition planning for 2025), I got a graduated education in network recruiting. When I consulted the literature on recruiting (which I did after the fact in preparing for this book), I found that the practice and theory largely agree. There is a science and art to collecting a useful crowd.

Building a Network

In the 1976 film *Network*, a has-been broadcaster garners instant attention by announcing he will commit suicide on his next weekly broadcast. That draws a crowd. He becomes a mad, mass media messiah, unleashing his power in weird and unpredictable ways. Long before the internet was a thing, this dark comedy forecast what the power of mass networking could accomplish. That said, there is nothing in this film (no surprise, it's fiction) to be learned about serious networking in the real world. National security efforts need something more dependable and predictable than by the off chance of happenstance of a madman collecting a crowd.

Mission First

For starters, here is a blinding flash of the obvious. Before recruiting people for a network, figure out what the network is for. Have a purpose. This is not throwaway trivial advice. This is important. There is a big difference between people who jump online to see what is up and those hungry, determined, self-promoters who ravage Instagram until they become a global influencer with their own line of cosmetics. They are driven by different goals, and the goals shape effective action online and offline.

Let's go back to the example of Russian disinformation. Although I have my issues with the efficiency and effectiveness of Moscow's meddling, there are parts they get right.

The Russian government doesn't try to influence people just because the former is evil. There is method in their malicious actions. They know what they want to accomplish. They want to dominate Europe like an MMA fighter rules the cage. Russia has long seen the key to its ultimate security through establishing a hard sphere of influence around the nation's frontier, with dominating influence over the political, economic, and military activities of neighboring countries. Not surprisingly, a significant amount of their effort is focused on the Russian diaspora in the surrounding countries, all of them either former Soviet states or Warsaw Pact nations that had gained significant Russian-speaking populations after World War II.

In many of these countries, both Russian-language broadcast and social media are prevalent, as are corresponding real-life networks of Russian activities from dens of spies and social clubs to criminal cabals. Some of these recruiting efforts have real consequences. Let's go back to the tiny country of Moldova as an example.

Most Moldovans are bilingual and speak Russian. The ethnic Russians in Moldova speak Russian as their native language. At the height of the COVID-19 epidemic in 2021, Moldova had one of the lowest vaccination rates in Europe despite having doses on hand and the infrastructure to distribute them. The reason, in part, is Russian disinformation, spread by real-world networks like the Orthodox Church as well as online social forums.[6] According to one study of postings on Facebook, the Russian narrative is about "combating/distorting the efficacy of Western-made vaccines and amplifying their side-effects; citing controversial/polarizing public figures whose opinions are aligned with Sputnik's; criticize the EU [European Union]/US response to the pandemic, in comparison with Russia's response."[7] The Russians aren't doing all this activity because they are interested in Moldovan public health. In fact, Russian disinformation is literally dissuading vulnerable people from taking the vaccine and putting their health at risk. The Russians have a goal. The present Moldovan government is staunchly anti-Moscow. The Russian objective is to sow distrust of the government, undermine the economy, and increase opportunities for Moscow to extend its influence by shepherding in a pro-Russian regime through harnessing the country's Russian-speaking population. The point is, the recruiting for this disinformation operation is purpose suited. Moscow is interested in people who will help shift public opinion in Moscow's favor. Russian speakers are a logical target.

So, step one in step one is not recruiting anyone and everyone. Step one starts with defining a mission for the network so you can be purpose-driven in building a purpose-built social network.

Let's caveat that.

In network warfare, you don't necessarily even have to have a plan at the beginning. In fact, sometimes it's better to build the network before you need it. After all, no one waits for a fire before they assemble a fire department. Setting up the hook and ladder squad first works because people understand the purpose. They know why they need a fire department and what they want to accomplish with one even if they don't know exactly when the alarm bells will ring.

Furthermore, since these are human networks, not automated bot nets, their mission and purpose might be dynamic and change over time. Just like military leaders start with a plan and then adjust as adversity or opportunities present themselves, a human network starts with a purpose and then can change as well. Here is yet another military analogy. When the Allies hit the sands at Normandy, they had a different plan for getting off Omaha Beach, but when it turned out that failed (they couldn't break out through the natural beach exits that were blocked by the Germans), US troops on the ground changed their mission and sneaked up the bluff to outflank the German positions. They adapted. Social networks can do the same. Adversity, or opportunity, can lead networks to morph into something beyond the original intent for social media action. So be prepared to adapt. These shifts happen in the virtual world. Several years ago, I was asked to brief the concepts in *Wiki at War* to an assembly of military attachés who the Defense Department had taken on a tour of Silicon Valley. The best part of the trip was (to be honest) not my speech. The best part was that I got to hang out with the attachés for the rest of the week and meet the glitterati of the tech world (including Vint Cerf who helped invent the internet). One of the activities was a trip to the Google campus. I will never forget one of the comments of a Google executive: "Right now, our main research effort is on figuring out how people use our search engine." That was a bit of a shocker given that a popular book on Google featured the company's organizing principle—"Know what business you are in."[8] Yet, here was a multibillion-dollar company that didn't fully understand how or why people used their products. To be fair, the comment was perfectly understandable. The founders, Sergey Brin and Larry Page, built the company to do one thing, but then as technologies, their capabilities, and the opportunities evolved, the company had to change to meet challenges and take advantage of new prospects.

So the point is that it is crucial to start with a mission while recognizing this is not a linear and static action. In a dynamic network, all the activities get revisited to make the network become more effective, powerful, and dominant. As for determining the mission, this takes us back to our earlier discussion. The mission of a network organized to impact a national security issue ought to be about—wait for it—national security. If it is not, in the end, the network won't be perceived as credible. Take, for example, the Project on Government Oversight (POGO), which manages a project called the Center for Defense Information, whose mission is to "secure far more effective and ethical military forces at significantly lower cost." The people at POGO are no doubt well-meaning, but they and their major funders, including the left-leaning Open Society Foundations, are not really interested in defense issues. They are interested in cutting defense and spending the money on other stuff. Although they have an active social media presence, they don't really attract engagement with serious national security people in Washington for one simple reason: those people know POGO is not serious about national security issues.

In some cases, networks might intentionally mask their interests, as in the example of Russian disinformation. Still, if the purpose of the network is not linked to national security issues with realistic missions, they won't deliver outcomes that intentionally affect national security. It's just that simple. Contrast, for example, Russian disinformation efforts in places like Moldova (where they have real impact) to the United States. According to one estimate, about 20 percent of Russian disinformation is in English. Since 2016, it is estimated the Russians have spent well over $100 million on these efforts. That's not an insignificant chunk of change. For what? Even though there was a lot frenetic reporting on Russian election-meddling and so forth, there is, in fact, scant evidence of significant direct influence. For all the effort Moscow invests, one major Russian effort, the Strategic Culture Foundation (SCF, founded in 2005), after fifteen years achieved the most miniscule social media presence in English. In 2020, the SCF had a little over 28,000 followers on Facebook, fewer than 500 on YouTube, and a dormant Twitter (now X) account.[9] Compared to the disruptions in Europe, this amounts to a pathetic peep in social media mass communication.

What's the difference between Moscow meddling in places like Moldova and Michigan? It is not enough just to have a mission. Get the mission right. Crafting a national security social network should link to a national security purpose with suitable, feasible, and acceptable tasks. (1) Suitable—the actions of the network should lead to the desired outcome. (2) Feasible—there should be a reasonable expectation that the task could be carried out. (3) Acceptable—the means and the outcome should be agreeable. Our transition support effort is an example. We were trying to impact a new president's policies. That was suitable. We were doing it through the transition process. That was feasible. We followed the law. That was acceptable. Our mission checked all the boxes. This is foundational to a successful recruiting effort. Let me explain why.

Messaging Mission

Having a mission is a start, but that does no good unless you can use the purpose of the group as a recruiting tool. That means there must be a capacity to deliver a recruiting pitch to reach the individuals you are looking for (recruiting) as opposed to those you just might attract (advertising). Whether through a post or an "elevator pitch," that short delivery in a quick conversation that takes as much time as riding one floor in an elevator (before presumably your target might bail and escape the elevator on the next floor), or any other means of recruitment, there must be a way to engage the networkers you want. An effective recruitment effort ought to contain three elements. The message should be credible, understandable, and actionable.

Let's walk through each component of the recruiting message.

Credible. In building a purpose-built network, the purpose (answering the question, why should I care?) matters. To recruit for our defense transition effort, the case had to be made that there was a problem to solve—to keep the military from going hollow. That wasn't a heavy lift. In 2012, there were already signs that the military was being run ragged and the administration was underinvesting in defense. Having Talent also helped with credibility; he had served on a nonpartisan congressionally chartered independent panel assessing US military capabilities. Furthermore, all the organizations that partnered with us to pull this together (including other think tanks) had reputations for doing serious defense work. If they were recognized as groups that took this seriously, then prospective members would as well.

Understandable. The ability to communicate matters (answering the question, do you understand what I am saying?). In the case of the transition network, this was not a difficult challenge either. If there is one thing that marks a successful politician, it's the ability to connect with people. In 2012, Jim was a master at helping us hone the defense message. During the 2016 election, the figure that emerged as the peace-through-strength candidate had his own way of delivering an understandable message. He called it "America First." Critics often attacked the motto as divisive. I happened to meet with the candidate and his team not long after he gave his first "America First" speech. Since I wasn't there as a campaign adviser, I didn't want to offer any suggestions on whether I thought it was a good motto or not. After the meeting, however, I strolled over to some of his staff whom I knew and asked out of pointed curiosity, "You guys know where 'America First' came from, right?" America First was the name of the organization that led the antiwar movement in the United States right up to the attack on Pearl Harbor. One replied, "We know, but we think most of the people that will vote for us have never heard of America First. Anyway, it is not what we mean. And, the people that will vote for us, they will know what we mean." They did have a point. After all, the candidate won.[10]

Actionable. A purposeful network delivers outcomes (answering the question, what can I achieve if I join this network?). The promise of action must be implicit up front. In the case of our transition network, we could point to the 2010 law. The presidential candidate was going to need a transition team that would prepare the way to the White House. I did not work on the 2012 transition, though I knew several folks who did and supported them as best I could with the resources of our network. In 2016, the day the transition team set up and started work out of a commercial building on Pennsylvania Avenue in space rented by the government General Services Administration, I went over to meet the national security team to offer assistance. A few days later, I got asked to join the transition team. Many others from our network contributed as well. We organized a similar support effort in 2020 and 2024. Not surprising, based on what we did in 2016, finding folks for the networks was not a problem.

Recruiting

With a mission and message, there is a solid foundation for recruiting a purpose-built social network. Now comes the hard part. Mission and message scope the effort a good bit—they define the crowd you are going after. The Russian disinformation example is a good case in point. Their most concerted efforts are targeted at Russian-speaking ethnic minorities in central Europe. They were the crowd most likely to be most instrumental in sowing discord for the Russians. Specific recruiting tactics must be tailored to the task at hand.

There are two crucial factors that impact not just the act of recruiting but also recruiting those who are most likely to be an active and productive member of a network. As one study that looked at recruiting social activists found, both the character of the recruiter and the characteristics of those recruited impacted the efficacy of the recruiting effort.[11]

The easiest question to answer is, who to recruit? If you understand the mission, you want networkers with the skills, knowledge, and attributes to contribute to mission accomplishment. To illustrate, let me offer the example of Brett Schaefer.

Brett has worked with me for years. He is one the world's leading experts on the mechanics of international and foreign affairs—how diplomacy gets done. One of the issues we wanted to tackle in transition was the operations of the State Department. It had been decades since there was any serious introspection and reform of US diplomacy. During that time, a lot of how the world turns had changed. We felt the State Department needed rethinking to be relevant as much as the Defense Department needed an influx of investment to prevent the armed forces from going hollow. We needed an Energizer bunny with the skills, knowledge, and attributes to lead this rethink effort.

Skills are set abilities to do things. Brett is a skilled writer and researcher.

Knowledge is mastering critical information that facilitates informed judgment. Brett is a no-kidding encyclopedia on how the State Department works.

Attributes are qualities and characteristics, having the demeanor, maturity, and temperament for the task. Brett is the most hardworking, honest, and selfless person I know (and one of the smartest)—a person people can trust and depend on.

It is also worth mentioning that Brett's approach to problem-solving is linear (the importance of linear and nonlinear thinking was described in chapter 1). He goes from a to b to c until the problem is solved. My approach is nonlinear; I jump around problems like a marauding pirate until striking gold or being hung on the yard arm. Together, on a big complex problem like State Department reform, we made for a pretty good combination, the kind of linear-nonlinear skill set that is most impactful in networking operations. Here was a case where it worked in real life. I threw Brett at the problem and Brett took it from there. After he did all the hard work, I connected him with the people who could put his ideas into practice.

Schaefer had the perfect combination of skills, knowledge, and attributes to lead our State Department transition reform effort. He produced a brilliant paper on the topic. Although he was not formally on the presidential transition team, they relied a lot on his advice. The reforms that Brett pressed were central to the agenda of the new secretary of state.[12] I know this for a fact because the guy who helped organize the reform process for the secretary was part of our network—and he told me so. Living proof: If you recruit people with the talents you need to get the job done, the odds of the job getting done go up.

The harder question to answer is, how to recruit? The closer the network is to operating on conversation mode, the more essential it is to handpick the participants because the success of the network rests heavily on the interaction between them, the skills, knowledge, and attributes they bring to the cause and their ability to collaborate and generate action. In 2012, for instance, we had a core leadership of fewer than a half-dozen people. We had all worked in the national security space for years. During that time, we had all on and off collaborated at one time or another and certainly knew one another well. A common bond is that we all knew, had worked with, and trusted Jim Talent.

The more the network resembles broadcast mode, on the other hand, the more recruiting is about looking for a type of person than a specific person. For instance, one of the interesting dynamics of the 2016 and 2020 campaigns was the nature of the group coming together to support the candidate and then president.

Normally, Washington is ruled by tribes. Leaders that rise to power usually have an extended political family of folks that they have connected and worked with through the years that form the nucleus of their administration. Many of FDR's New Deal colleagues, for instance, came from a circle of friends going back to his days in the Wilson administration, neighbors who lived around DuPont Circle. JFK famously drew from the "Georgetown set" and his Ivy League connections. Occasionally, however, an outsider comes to the Oval Office without a tribe. Then newly elected presidents must scramble to build one. That was partially true for Jimmy Carter, the peanut-farming governor from Georgia. True for Ronald Reagan, the former Hollywood actor. And then some for Trump, the real estate mogul, builder, and reality TV star from New York who was less interested in credentials and Washington experience than people who could advance ideas in his America First agenda. For them, building a trusted network required looking outside the usual suspects of Washington experts, a recruiting task that was as different as could be imagined from building a support team for more traditional insiders like a Bush or a Biden.

Whoever the team must rally around, if you are building a big team, you need to think big. The best solution is to find a networker to build a network. People are more likely to join with those they know and trust. Networkers with the talents to recruit people are also likely to recruit bigger and better networks. The person who knows more people or who has the capacity to influence more people is more likely to recruit the network you need.

Let's take an example from history. The original America First movement is a good one. The committee that guided the movement opposing US entry into World War II wasn't formed until September 1940. The movement's rocket to prominence was triggered in part by the fact that one of its most high-profile leaders was the famed aviator Charles Lindbergh, perhaps the most well-known, respected, and trusted of public figures in America. At the height of the organization's influence, the committee drew a packed house for its rally in Madison Square Garden. National polls showed 80 percent of Americans agreed the United States should stay out of the war. This group had real impact. After Pearl Harbor, Lindbergh led in the opposite direction. The organization endorsed the war effort. Lindbergh and other leaders volunteered for military service. The lesson is, in networks as in life, strong leaders can lead.

In 2012, Jim Talent pretty much knew "who was who" in the Washington defense community. Recruiting wasn't a real challenge. After building that network, our contact lists only grew. By 2020, our network of names was extended enough that even reaching people outside the usual Washington cast of characters wasn't a real challenge. Among others, reliable colleagues, like David Shedd who had national connections with talented people not just in the intelligence community but also in the private sector, helped us spread our reach.

Uber-networkers also tend to be in other networks. They can cross over from one to recruit in another. Use trusted networks to recruit for networks. For example, when Sara Carter started Darkwire, her guerilla media project, she needed to recruit young people to train as investigative journalists. No problem. We enlisted other networks of students and young professionals to help find the right ones for her program. Cross-linking networks to build teams is an effective practice.

Networkers may also be influencers and tap into social media to increase their influence. There is a reason why I stayed close to Sara Carter all these years: in addition to her being a genuinely nice person, she has a powerful social media presence. When her roles in conversation mode networks and broadcast networks are connected, they can be seriously impactful.

Of course, networks can recruit also using the capabilities of social media. There are all kinds of tools and tactics, like clickbait (hot links often with provocative or sensational titles or pictures intended to attract users to click on and follow a link). There is a vast literature on how to use these. Unfortunately, I don't think it is particularly useful or relevant. Here is why. Most of this recruiting activity is focused on delivering revenue to the recruiters, or customers, or suckers to be fleeced by online criminals, or masses to be manipulated. They are not structured to build action networks. Better models look more like what we describe. They look more like how community activists command influence or extremist groups recruit extremists. That's because the goal for them is not just building networks but networks that trigger action.

As an example, look at the literature on community organizing. The fundamental goal is to recruit participants who share a belief that in their collective action they can produce specific outcomes or change. Successful recruiting will come from practices and activities that deliver that kind of result.[13] The point is, there is a significant difference between advertising on social media and recruiting for an action network. Advertising appeals to what people want. Recruiting draws people to what they need.

Social media platforms can be superb for recruiting (if the platforms and access to the platforms and devices is available to the people you need to recruit). There is such a plethora of apps with so many different formats and capabilities from chat features to video that the task is to pick the right one for your mission and your message. That means doing your homework, understanding the platforms, their capabilities, who is on them, and what they use them for. We discussed some of the differences and features in chapter 1. LinkedIn, for example, is recognized as a professional networking site, popular with college students and graduates—a more affluent and educated crowd looking for professional relationships and activities.[14] Social activists are more likely to use other platforms. Tumblr, for instance, is favored for social activism by high school, college, and graduate students as a convenient platform for posting commentary. Another factor that will influence the choice of platform is whether your network is operating in broadcast or conversation mode. Broadcast suits applications like X where potentially any of two hundred million users might see a particular tweet. Conversation mode is better suited for platforms with more intimate and targeted communications like Snapchat or WhatsApp.

Supplemental technologies can also play a role in crafting recruiting strategies. Conventional web tools, however, are not optimized for this. There are special tools and new emerging capabilities that can prove useful, some that might even prove game changers. One technology, for instance, is "link analysis" software. Link analysis uses data to evaluate relationships and connections between nodes. Link analysis software is used for all kinds of activities from research and marketing to intelligence collection and mapping terrorist networks.[15] Link analysis is only one instrument in a family of data-mining technologies. If you are searching for an informal network or looking to build one and have access to the right databases and the software, these technologies may help. More on technology later (chapter 14). Still, although technical tools can assist, at root impactful networks are about people. Look at recruiting as a people activity, not a technology problem. If you must turn to technology for help, look for technology to supplement, not supplant, the human role.

Finally, reality matters. On October 7, 2023, Hamas attacked Israel triggering a protracted regional conflict. Preceding the fighting, I had spent over two years trying to recruit a coalition of Jewish and other policy groups to form a coalition for combating anti-Semitism and supporting US bilateral relations with Israel, making scant progress. After the bloody assault on October 7, in less than three weeks, over thirty organizations voluntarily joined our network. Groups suddenly recognized there was a need to band together.

Offense and Defense

As discussed in chapter 2, tactics are about offense and defense. While a networker can recruit a network for the purpose of taking offensive actions, others can exploit that effort to turn recruiting against the network. One tactic is to infiltrate and disrupt or exploit the network. Another is to delegitimize the network, so either folks don't want to join or undermine the integrity of the group. More on that in later chapters, but the point is, when recruiting for a network, think about how to protect the resiliency and credibility of the network against malicious actions to undermine your efforts.

Protecting your network when it comes to the task of recruiting (as well as the other activities *Digital Dominance* will cover) is mostly about mitigating risk. Risk is measured by looking at the combination of (1) threats—what might try to mess with the network and its operations, (2) vulnerability—the susceptibility of the network to disruption, and (3) criticality—how vital is it to protect the network.

Threat. A threat analysis looks at both the intent and capability of an actor that might come after your network. It makes a difference if you are worried about interference from a hacktivist or being attacked by an armada of billions of Russian bots.

Vulnerability. Vulnerability assessments evaluate how risky operations are if a threat comes after your network. For instance, if your social media activities are supported by a dependable service provider and your network follows responsible cybersecurity practices, networking activities are probably safeguarded against the most common malicious cyber activity. On the other hand, if your privacy settings are set to "none" and networkers are clicking on promises from a Nigerian prince to share his fortune, you are likely doomed.

Criticality. Criticality is a measure of how vital the network is and how important it is to prevent disruptions. Many cybercriminals, for example, don't worry much if their operations are disrupted. They don't sweat when YouTube blocks their bots. They'll just send another bot army into the breach. They have billions of them. If, in contrast, your network is collecting real-time intelligence on the battlefield, losing that capability in the middle of a firefight would likely be much more of a worry.

An assessment of all three factors informs on what the important risks are that must be mitigated. Then you can start thinking about the appropriate measures to address them. This process holds not just for recruiting but every phase of building and running a purpose-built network. It's a topic we will come back to when we cover each step.

So how does mitigation work in practice? Returning to the presidential transition support network as an exemplar, let's look at how we handled the challenge of protecting the integrity of our recruiting effort.

First, we had an open tent for anyone who wanted to participate. The only caveat was a shared commitment to peace through strength as the guiding principle of defense planning. This requirement ensured that we were building a network of like-minded folks committed to joint action, who could work together in trust and confidence. We didn't recruit people who would be a problem.

Second, we didn't formally vet folks, but individuals that joined were recommended by individuals who worked with us. That was how we built out a network of trust.

Third, we decided early on to be completely transparent. We didn't have any secrets, let alone secret handshakes. If there is nothing to hide, there was nothing to compromise. This meant that we didn't do much more than what one normally does to safeguard online activities from malicious actors. The basic rule was, "Never put anything in writing that you would be embarrassed about if it showed up on the front page of the *New York Times*." This was our way of mitigating risks to the network.

There is no fixed formula for protecting network activity. The measures that might suit for a network depends on the network. That prompts another reminder—why it is good to start with a mission and a message. Knowing the nature of the tasks you will be getting into also prompts thinking about who might come after you. Once you do that and after you figure out how to address and mitigate risks, you can recruit with confidence. There will be more on protecting your network in the next chapter, but this is enough to get started.

Now you are ready to launch. If you have a mission and a message, follow good recruiting practices, exploit the abilities of skilled networkers and social media platforms, you can start to assemble the army you will need to dominate the digital high ground.

What's Next?

Speaking of recruiting, years ago I recall an organization that spent a bunch of money on advertising to recruit supporters. It worked. They got hundreds of thousands of new members. They all sent checks or credit card numbers. Only problem? After a year, many drifted off and did not renew their memberships. Why? They didn't feel like they had a reason to stick around. Worse, the cost of the advertising exceeded the income from membership dues. In the end, a lot of resources were expended without much to show for it. Solving that problem in a network is the next step in building a dominating place in social media warfare. How do you keep and add to the network you have recruited? That's covered in the next chapter.

4

AN OFFER YOU CAN'T REFUSE

Everybody knows Bridgett Wagner.

If you don't know Bridgett, you probably know someone who does, or know someone who knows someone. Trust me. It's like the parlor game six degrees of Kevin Bacon, where you must connect Bacon to another actor by naming films they have been in with another actor until you make a connection. This game is an illustration of real science, how the interconnectedness of the modern world is a real thing, called the small world phenomenon. Research on the topic was pioneered by a sociologist named Stanley Milgram, an influential scholar we talked a lot about in *Wiki at War*, one of the pioneers in our understanding of how social networks work. Wagner would give Milgram pause. She is an oddity of science. I bet she can connect with anyone on way fewer than six people. Bridgett Wagner is a networking diva.

Who would have thought that a girl from little ol' Richardson, Texas, with a bachelor's degree in economics from the University of Dallas, would grow up to be one of the most connected people in America? Probably the guy that hired her.

Wagner got a job out of college as a research assistant and speechwriter, the same start probably a billion other young people get in Washington, DC. But rather than moving from a starter position to a normal career like most people, she took a different path. Rather than put Bridgett into a job like everybody else, they created a job to suit her talents.

It didn't take long for folks to see that Bridgett's real skill wasn't filing. She was a people person. And more than just a person that liked people, she was someone who had an intuitive sense of which people should be connected to which people. After talking with someone for sixty seconds, she would start doing this mental jujitsu thing—whom should I connect this person to? She is genetically predisposed to put people together. If Bridgett were a matchmaker, she would have eclipsed the mass marriages of Sun Myung Moon's Unification Church many times over.

Once her boss figured out her true superpower, he launched her off with the mission to build coalitions. It was like unleashing Godzilla to trample Tokyo.

Bridgett has spent her entire professional career, well over forty years at last count, building coalitions across the political landscape abroad and in the United States, in Washington, in state governments, with nongovernmental organizations, and think tanks, and activist groups, university students, and the private sector. She is a Swiss Army knife of networking—a thought leader, peacemaker, negotiator, speaker, motivator, encourager, adviser, coordinator, and counselor building a network of institutions and issue-based organizations in America and around the world. You are likely to hear someone say, "I know Bridgett Wagner" in Lisbon, Portugal, as it is in Loma Linda, California. In fact, on my first visit to Baku, Azerbaijan, in 2022, one of the first questions a stranger asked me when they found out where I worked was, "Do you know Bridgett Wagner?"

Like many master networkers, her networking led to participating in other organizations and other networks. For example, in addition to her official coalition-building day job, she is an adviser, board member, and participant in all kinds of organizations including the Steamboat Institute, Docs4Patient Care Foundation, Talent Market, the Antigua Forum, the National Civic Arts Society, the Henry Luce Foundation, the Mont Pelerin Society, the Philadelphia Society, and more. She is valued not because she is a big donor or a big influential thinker, but because when you enlist Bridgett Wagner in your cause, that comes with the global connectivity she provides. It is her ability to plug the world together that makes her a master influencer without influence.

One of her initiatives, the annual Resource Bank meeting on nonprofits, is a gathering of hundreds of groups from across the nation and around the world to meet and trade ideas, initiatives, practices, lessons learned, and network. It is worth noting here because Resource Bank gets to the real secret ingredient of Bridgett's success as a coalition builder. It is more than just connecting with people. Year in and year out, people pay their way to come back to Resource Bank for an annual frenzy of networking. Why? Why do the people Bridgett recruits stay recruited?

Unlocking the Code

I do a lot of work with the international community in Washington, D.C., paying particular attention to the embassies of small- and medium-size countries that are strategically relevant to US interests. The reason why is simple. They pay attention. Great powers like to rub shoulders with great powers. If Beijing calls the White House, the White House is going to answer. Lesser nations have a harder time getting attention, like the little kid in the back of the room furiously waving their hand to be noticed by the teacher. Their embassies have smaller staffs, no pricey lobbyists, and few powerful friends. They are small markets for US businesses. Their diaspora in the United States doesn't make up a powerful voting bloc. They don't buy a ton of US weapons. Yet, for one reason or another, where they are located or what they do, they matter to American foreign policy, and building stronger bilateral relations with these countries is in the United States' interest. Often, official Washington doesn't have the time to give them the attention they deserve. Unofficial Washington often steps in to pick up the slack. It is, after all, in our interest to serve the national interest. The embassies are grateful for the engagement.

It is always a challenge, however, to find ways to be useful. One step we can't take is to represent a government or its embassy. That gets complicated legally very quickly. Lobbying in Washington comes with all kinds of restrictions and reporting. Lobbying for another country comes with additional requirements under the Foreign Agents Registration Act. A further complication for me is that my think tank has a policy that prohibits influence, donations, or grants from any government, either ours or others. All these strictures make sense; they ensure our independence from outside control. In short, we can't do stuff for the embassies, but we can do stuff with the embassies. What I found over the years is that the stuff the embassies are most interested in is the stuff that most benefits—the embassies (I know, this isn't rocket science).

A few years ago, the head of our think tank suggested we invite representatives of the embassies to Bridgett's Resource Bank. Since it was my boss's idea, I liked it. No, I honestly really did like it. Here is why. The United States is a huge and daunting country, and the job of embassies was to understand all of it, not just what was going on in Washington. In fact, many of the issues embassies are most interested in are often not diplomacy but economic opportunity, education partnerships, innovations in governance and public services, and trends in American politics and culture, all kinds of activities that happen more outside than inside the Beltway. Small embassies don't have consulates spread all over the country or expatriate communities in all fifty states. Why Resource Bank? At Resource Bank, they had a one-stop shop to meet and engage people from all over the country, people who were interested in many of the same kinds of things the embassies were interested in. So we invited ambassadors.

It worked. Here is why. I might have been able to recruit embassies to come to Bridgett's networking orgy because I am a smooth talker, but they stayed because they got something out of the experience that they valued. This is an example of the second step in building a resilient, active, and action-oriented network. After recruiting them, give people a good reason to stay engaged. A successful network will deliver a value proposition to its users.

This was when I decoded the magic behind Bridgett's networking abilities. She had an intuitive ability to envision how the people she connected might help one another. Bridgett built relationships based on satisfying the value proposition. That was why she succeeded at networking more than most. The reason you may personally not know Bridgett Wagner (though someone you are connected to inevitably does) is that what Bridgett Wagner does has never been about Bridgett Wagner. It has always been about the mission of making connections of value. Her reputation is built on that simple, selfless act. So even if you really don't know Bridgett and really don't know anyone who knows Bridgett, odds are you may have been helped by someone she has helped.

Value of the Value Proposition

One of my all-time favorite films includes one of the most iconic lines in cinema history. When Don Vito Corleone is asked how he is going to get what he wants in *The Godfather* (1972), the don replies, "I am going to make him an offer he can't refuse." Keeping folks engaged in a serious national security social networking effort requires that kind of compelling offer.

The Bridgett Wagner way of networking is more relevant than it might seem at first blush—even for national security confrontations. Leadership is a relationship between the leader and the led, requiring both communicating and motivating. This dynamic is present as much in the military as anywhere. Although most think of the American armed forces as a hierarchical, dictatorial organization where the troops just do as they are told, the reality is very different. There is no denying that sometimes soldiers do what they are ordered to because they don't want to be thrown in the brig, or demoted, fined, or tossed out of the military. That, in my twenty-five years of military experience, however, was rarely why soldiers followed commands. The value proposition played a much bigger role.

Here is an example. Prior to Operation Desert Storm, a young officer came to me in near tears. He had been ordered to deploy with one of the units going to the war. He was a new father, uncertain about facing the prospects of real war. My advice was, he joined the military, he took an oath, honor it.

I had known and worked with him for a long time, I knew he was an honorable person, and honorable people honor their commitments. He went. He served. He also came back and saw his child grow to adulthood. The point was that despite his doubts and misgivings, honor was important to him. It had so much value for him that he was willing to risk his life fighting beside others to keep it intact. Value matters.

There is a vast literature about why soldiers fight, covering every aspect from patriotism to small-unit cohesion, but they all agree it has less to do with fighting because they are told to fight than fighting for a reason—something they value.[1] People need a good reason to be shot at and not just call it a day and walk away.

In the military, value propositions play a role not just in leading soldiers but also in dealing with the world of a battle. Remember the group mentioned in the prologue, Spirit of America? Their founder, Jim Hake, recalled an incident that illustrates the value proposition well.[2] This is actually the story of two Jims, two Jims who could not have been more different. Commissioned in 1972, Jim Mattis was a grizzled Marine Corps "lifer" who had fought two wars in Iraq and one in Afghanistan (yes, the Jim Mattis who would later wind up as secretary of defense, call sign Chaos). Hake was a Silicon Valley entrepreneur who had cofounded and sold one of the first internet media companies, then cofounded another firm that paved the way for today's mobile apps. Mattis was a no-nonsense master of military virtues. Hake was an experimental business builder.

What united the two was 9/11. The attacks had profoundly disturbed Hake, and he shared the general's view that "we are in the middle of a violent global argument between the voices of intimidation and inspiration." Both men anticipated that US troops would face difficult obstacles in the Sunni Triangle in Iraq. Mattis's marines were first-class fighters, but his mission required much more than combat. Winning the battle of competing values would require building community services, working markets, and a vibrant civil society in the villages in his area of responsibility.

Hake told Mattis his idea: He would create an entrepreneurial, philanthropic organization that would help US military forces achieve their noncombat objectives. It would raise money online and hire retired military men who would accompany active troops and diplomats into disputed neighborhoods to help them identify needed goods and services. After an ultraquick review process, the nonprofit would deliver the desired commodities for distribution by the troops. It would be a decentralized, fast, targeted effort, focused on solving problems and building goodwill. Structured like a well-run business, this charitable effort could speed its assistance to the battlefield faster than any cumbersome government entity could. In more ways than one, it would represent the spirit of America, which is what Hake called the group.

Mattis left for Iraq with Hake's number on his speed dial and put it to good use. One of the more unusual calls came in 2006, at the height of what was called the Anbar Awakening. The local tribes were beginning to reject al-Qaeda and Mattis knew it was a moment brimming with opportunity. He had a chance to win the support of the local community, but he needed a powerful gesture to help bridge the distrust between his commanders and the Sunni chiefs. He knew that in the local culture, a sword has deep symbolic meaning. When a sword is exchanged, disagreements and retribution are put aside. Accepting a sword is a sign of trust and friendship. It was exactly what he needed.

Mattis had plenty of armor, but swords were another matter. He couldn't use taxpayer funds to buy gifts, let alone $600 marine officer ceremonial swords. But he knew someone who could get them if asked. He contacted Hake, who raised the funds through Spirit of America to buy and deliver a dozen swords. Mattis's successful effort to recruit tribal chiefs to help end the "Sunni insurgency" that had devastated the country after the Iraq War was due in part because he could build a relationship with the chiefs, giving them something they valued: not only peace and security for their tribes but the symbolic bond of a marine sword.

If the value proposition works in matters of life and death to forge action-oriented activities, then it ought to come as no surprise that its power in sustaining dominating social networks would be just as profound, particularly in matters of national security.

The real experts in value propositions are in the private sector. I know in previous chapters I spent some ink denigrating commercial web practices, tools, and advertising, but when it comes to this task in social networking, they are the masters. Fundamentally, marketers understand the human dynamics of a value proposition as well as anyone.[3]

There are some caveats to applying what marketers do to the practice of social networking for national security. For marketers, the value proposition is a sales pitch for a service or product. That isn't the only value of a value proposition or the only form a value proposition can take. For networkers, the purpose of the proposition is to keep networkers in the network. For national security networkers, although the mission and message may be enough to draw them, different propositions might be needed to continue to demonstrate the utility of the network. These value propositions also might take many different forms. They might change over time. More than one value proposition may be put in play to keep networkers networking. Think of a value proposition more like a menu than a line-item request.

Furthermore, unlike linear marketing, which looks at an explicit link between the money spent and the revenue received, the return on investment (ROI), an effective value proposition for other networks, may manifest itself in less tangible ways. Trying to measure an ROI that can't be codified in dollars and cents gets back to the challenge raised in an earlier chapter: How do you measure the effectiveness of social networking when the value of the networking is not an immediate observable concrete action?

Look at the cases of WeChat and WhatsApp, some of the fastest growing social networking tools on the planet. One study on these services identified a complex array of factors that impacted the customer value proposition and the acquisition strategy for attracting new customers, including market mission and values, competitive positioning, market access, and competitive differentiation. The bottom line of the study was a recommendation for WeChat that improving product performance without raising price was the key to attracting an expanded user base in China, a counterintuitive recommendation to what might have come from a more traditional ROI analysis.[4]

Building the Proposition

At its core, a powerful value proposition is about relationship building, forging a bond across a checkout counter or a virtual network. I once met an owner of one of the premier high-end hotel chains in the world. He explained the key to his phenomenal success. It wasn't just about acquiring customers. He lectured, once you get a customer, never let them go, never give them a reason not to come back. In their hotel, every employee was empowered to spend thousands of dollars to fix a problem raised by a guest. They wanted everyone to know the guest would always be taken care of—that was the real worth they offered. This was a business model built on a value proposition that was not a sales pitch but a demonstration of caring.

The value proposition establishes the belief that the customers' wants, needs, or fears are addressed by products or services the provider delivers. In the case of a high-end hotel, the value wasn't the price of the rooms for rent; it was that the customer knew they were being taken care of—always. For affluent customers who could afford to stay anywhere they want or pay any rate, that sense of customer service was more valuable than getting an Expedia good deal on a room rate. In short, what was good business was really understanding what their customers value, what keeps them coming back. That approach to crafting a value proposition is just as useful in social networking warfare.

In the marketing world, good marketing starts by understanding the product and the potential customers. It would make no sense, for example, for Motel 6 to sell itself like a Hilton. They might be able to "leave the lights on for you," but Motel 6 was never going to be able to credibly promise to meet Lady Gaga's every whim. A value proposition that doesn't create value the customer wants won't keep customers. In a similar manner, networks that don't deliver the value promised will lose the users they recruited.

Likewise, marketers must understand their market. It would, for example, make no sense for five-star hotels to market on Priceline—that's not where their customers look for a room. Likewise, networks that are looking for users with special skills, knowledge, and attributes have to know who those users are and how to find them.

Once you understand both sides of the relationship, it is time to craft a value proposition. There are two key parts to that. Part one: Know what the value is.

Take the example of the Flat Earth Society, founded in 1956. In the 1970s and 1980s, then-president Charles Kenneth Johnson became a minor celebrity by making the case (really, what the hell did Galileo know?) for the world being flat. At its height, his group had 3,500 members. Today, the flat earthers still flourish online with podcasts, videos, and websites. How is that possible? What possible value can come from a network arguing the world is flat? The answer is the society is providing a value its members value—and it's not evidence that the earth is flat. "Many people are willing to believe many ideas that are directly in contradiction to a dominant cultural narrative," concludes Eric Oliver, coauthor of a study on why people value conspiracy theories. "If you're faced with a minority viewpoint that is put forth in an intelligent, seemingly well-informed way, and when the proponents don't deviate from these strong opinions they have, they can be very influential," adds psychologist Karen Douglas. The value that members get from the society is a confirmation of their own judgment that thinking differently and independently on issues from everyone else is admirable.[5] Whether the Flat Earth Society can prove the earth is flat or not is irrelevant. What the society delivers is something much more important: a value proposition that satisfies its members. The lesson here is clear: To formulate a powerful value proposition, start with identifying a value that really matters.

Once you have clarified a value, formulate your proposition. The simplest and best advice from the marketing world is to make your value proposition explicit, clear, and delivered with evidence to back up your claims of value.

Here is a practical example of how that works from the national security networking world. After the 2020 election (my third presidential cycle of trying to build transition support for presidential teams), I had an epiphany. Why did we keep reinventing the wheel every four years? Why not just sustain the network until the next election cycle? Here was the problem. It was one thing to sell a value proposition for a network in an election year. People thought they might have a real opportunity to influence policy and personnel in a new administration. The situation, however, is a lot different when the election is four years off. There isn't much to do. No one is sure who the candidate might be and if those in the network would be at all interested in supporting them. It was anyone's guess that far out if any candidate had a prayer in getting elected. So what was the value of networking? We hit on one, and it proved effective.

One of the activities managed by Bridgett Wagner's coalitions team is a job bank that collates weekly lists of open jobs and helps connect job seekers with employers who need help finding employees. There are, of course, plenty of commercial job searching sites, like Indeed.com. Also, folks can hunt for their own jobs online through sites like LinkedIn. Bridgett's job bank, however, is more efficient. It narrows the field considerably between likely employers and employees. Furthermore, it includes the kinds of jobs the people in our network are most likely to want, such as congressional staff, nongovernmental organizations, think tanks, advocacy groups, and so forth.

One of the steps we started to take was to distribute the weekly job bank openings with our networkers, which included a fair number of people looking for jobs, helping other people look for jobs, or looking for people for their job openings. Network members valued this access. How do I know that? They tell me all the time. Even if they are not looking for a job, they appreciate that the network makes the effort to help others. The job listing is a clear demonstration that the network is there to provide a selfless service to others.

Getting the Proposition Up and Running

There are proven techniques for getting the most out of a value proposition. Let's run down some of them.

Test. It might make sense to try out your value proposition, before you roll the dice and put your credibility and an enormous amount of networking effort on the line. Test the value proposition with networkers you are trying to target. In our transition network, after we circulated job lists a few times, we quickly discovered the demand was high. To be honest, we had no idea of using the job listing as a value proposition; we thought we were just doing something nice to help people out. Once we realized the community came to depend on the announcements and value them, we started sending out the job announcements as a weekly service.

Manage. The value proposition is a tool to an end, not the end. Sometimes, it can be used to reshape the purpose of the network or create new networks. An influencer operating in broadcast mode could deliver a robust value proposition that helps grow a vast network, that might in turn be used as a recruiting base for other networks for other purposes. We drew from the transition network for our Afghan response group, for example, by soliciting members with a different value proposition—saving lives.

Match up. One of the inherent values of social networking for national security is networking, the value of bringing like-minded people together who might have interests and activities in common. In Washington, networking is considered one of the most valuable commodities on the market, better than a parking spot near the Capitol. Here is a small bit of proof. As Washington began to emerge from its post-COVID cocoon, we organized a reception for the members of our national security and foreign policy network. We were shocked by the turnout. Folks were anxious to get back out and about and connect and catch up. Virus be damned.

Monitor. One issue with a strong value proposition is that it might attract a lot of slacktivists but won't be useful in delivering action on the network's vision. This is the "Save Darfur" problem. An attractive value proposition that blows up your network numbers but doesn't deliver any outcomes isn't much help in networking for national security. For example, we did have folks who hung around our transition network just because they were interested in job leads or networking with other people. Addressing that challenge of free riders (networkers who just join for the benefits) is an issue to be addressed in the next chapter.

Offense and Defense

The value proposition, like every aspect of networking, can be weaponized for or against the network and its goals. As an offensive tool, a powerful value proposition can immensely increase the ability to not just recruit and retain members but also utilize the talents of networkers with the right set of skills, knowledge, and attributes to advance your goals. The best example here I can think of is the one mentioned at the start of the book—our Afghanistan response group. Our value proposition was that we were going to save lives and we were only interested in working with people who were going to save lives. That attracted support from the folks like a former national security adviser to can-do nongovernmental organizations like Spirit of America who secured access to all kinds of resources, senior officials in the United States, and influential players in foreign governments.

The value proposition can also be the focus of attack by a network's adversary. The center of gravity of the value proposition is credibility—do the members believe the network will deliver the value promised? Take the example the Taliban's efforts to convince the world that they would deliver good governance for the women of Afghanistan, and therefore the Taliban were worthy of support, engagement, and empowerment by the outside world. The credibility of that promise was demolished by reporters like Sara Carter and Lara Logan, influential voices like Ambassador Kelley Currie, and many others who, in a New York minute, undermined the Taliban claims with facts.

Conversely, a strong value proposition can also be a powerful defensive tool. If folks believe in the network, they are more likely to defend it or remain participants if the network suffers a reputational attack. Attacking reputations online is a common occurrence on social media. Take, for example, the Hollywood actor Jonah Hill. In 2009, he got into a furious Twitter feud with another well-known celebrity actor-director Jon Favreau. He then started posting weird, vulgar, and obnoxious comments about other Hollywood glitterati. People were getting really annoyed with Jonah Hill. The only problem: It wasn't Hill. To save his reputation, Hill raced to put in an appearance on the *Late Night Show* with David Letterman, where he announced, "I have never been on the website Twitter, nor will I ever be on the website Twitter." The Twitter scam was a fake account, the product of an inspiring actor and comedy writer.[6] Turned out, this was not a malicious attack, just a case of bad judgment by a guy who pretty much crushed any hopes he ever had of making a friend in Hollywood. Nevertheless, it was a real problem for Hill, as are reputational attacks on others that occur on the internet all the time. One way to help fend them off is to have a resilient, committed network to begin with. That is where a strong and credible value proposition can help.

Whether employing value propositions to build up a network or attacking them to drag a network down, operating in both broadcast and conversation mode can be helpful and having the capacity to link them together can provide a distinct networking advantage. Poor Jonah Hill learned to his chagrin what happens when someone else grabs the digital high ground ahead of you. His reputation was trash talked on Twitter until he counterattacked with an even bigger megaphone on broadcast TV to get his reputation back in the game. On the other hand, conversation mode is often ideally suited to strategic planning crisis management, plotting moves and countermoves that can be unleashed in bigger networks to establish dominating influence.

Okay, so let's recap. A value proposition is crucial for keeping people in your network and engaged. To put a proposition together, you need a clear understanding of what the network can do and the people you want to network with. Then figure out what your value proposition is and express it in explicit and clear terms delivered with evidence to back up your claims of value. Continue to test, refine, manage, and cultivate your value proposition after you deploy it. Finally, remember that your value proposition is part of your weapons arsenal for offensive and defensive action.

Let Slip the Dogs of War

Up to this point, the focus has been on building a weapon for social networking warfare. So far, *Digital Dominance* has covered how to recruit battalions of networkers to carry your fight to the enemy. *Digital Dominance* then addressed how to keep networkers around and engaged until you need them. What these last two chapters have delved into is the iceberg nature of powerful, dominating networks, all the activity that goes on behind the scene before the first post on Facebook, the first pic on Instagram, or the first shout-out on Yik Yak (a smartphone app that users share discussion threads within a five-mile radius). There is a lot of work in purposeful networking.

Now it is time to get down to business. National security networking is about doing things—assembling purpose-built networks to drive real outcomes.

Time to move on to how to unleash your social network on the world. That is the topic of the next chapter.

5

BECOMING FORREST GUMP

Terrell Chandler is an operator.

On her LinkedIn page, Terrell gives her philosophy of life: "It helps when you know how to do things, but it helps even more when you know how to treat people." That statement says everything about her. She has spent her whole professional life doing things—for others.

It would have been hard to guess that someone who started their career buried under the paper wars of medical services management in the rural Deep South was going to live a life of danger and adventure all over the world among real wars. But there you go.

A graduate of Arizona State University, Terrell served in the US Navy Reserves. That means she had a civilian job in real life but, if needed, could either be called up or volunteer for a deployment with the navy.

In 2013, the navy called. She was heading to Afghanistan.

True, Afghanistan is a landlocked country, but in war, it is all hands on deck. One of the assets in highest demand in the country was highly trained naval special forces known as sea, air, land teams (SEALs; we will meet more SEALs later in the book). In addition to their fighting skills, SEALs and other special operations forces like the army's Green Berets also work with indigenous military forces and populations, dealing with insurgent and terrorist threats. This is not too dissimilar from the kinds of activities Jim Mattis did in the Sunni Triangle cooperating with Iraqi tribal leaders.

Terrell was assigned to the Naval Special Warfare's Cultural Engagement Unit, a new capability the navy had organized to deal with the post-9/11 challenges the armed forces had encountered. The military found that in Muslim countries, many women were only comfortable or permitted to engage in public with other women. The program qualified women as combat support enablers, deploying them to support Navy SEALs on the battlefield in overseas contingency operations.

Chandler joined an elite group providing operational support alongside Navy SEALs and Green Berets, helping assist and coordinate with the local population as part of what was called a Cultural Support Team.

After ten years of knocking around with the navy, Chandler joined Spirit of America. Here is why.

The founder of Spirit of America, former software executive and business entrepreneur Jim Hake (whom we met in the last chapter), knew what business he was in—and it wasn't selling software.[1] Spirit of America was a new breed of humanitarian organization. Traditional humanitarian groups hold that nongovernmental organizations operating in conflict zones should stay scrupulously independent. Aid should never be distributed with political, economic, or military objectives in mind. The Spirit of America model, however, is intentionally different. "This is not conventional charity. It is not neutral. Everything is done in support of US troops," says Hake. This also meant that if Spirit of America was going to work, it was going to have to go to difficult places and figure out how to do hard things.

Since 2003, Spirit of America, following US forces and State Department officers into some of the toughest areas in the world, delivers private assistance intended to complement their work and advance US interests. Field personnel work alongside deployed troops to understand local conditions, identify high-priority needs, and decide what kinds of specialized aid could be brought in to help achieve security objectives. The nonprofit keeps a cadre of cultural and technical experts on call for whenever specialized advice is needed. It coordinates all its activities with US military and diplomatic officials.

To operate in the way Spirit of America operates, it needs operators. "The skills of our field representatives are key," notes manager Isaac Eagan. All of Spirit of America's field teams are US military veterans who previously served on the ground in Iraq or Afghanistan, giving them the experience, skill, and maturity to act safely in risky places. "They have to be able to assess and develop relationships, think about problems from an entrepreneurial perspective, leverage local resources, and above all, be flexible." Spirit of America runs on operators.

That's why Jim Hake hired the likes of Terrell Chandler. Not only was she a combat veteran, but Terrell was also an avid mountain climber, marathoner, and tandem skydiver. Add that up and it means she must be personally brave and not averse to doing crazy things. But more than lacking the sense-of-personal-safety gene, Terrell had a lot of experience organizing things in places that don't like to be organized. She managed Spirit of America's field operation team as deputy director for two years. In 2021, Terrell was named director of field operations.

Any one of Terrell's operations sounds like an improbable adventure novel. In 2019, for instance, she was in the remote Luang Prabang Province of Laos trying to convince people that milking a buffalo was not a crazy idea. She wrote in her blog, "As I noted in my email a few weeks ago, the US Embassy in Laos and Spirit of America have teamed up to support the Laos Buffalo Dairy as it provides opportunity and additional nutrition for the people." At first, the locals thought these foreigners were nuts. The crazy idea worked, as one of the newly established dairy farmers wrote in her diary, "Now, others rent their buffalo to the Dairy, and the buffalo are healthy and vaccinated. With the extra money they earn, people are able to do other things for their families." It was a small but incredibly powerful effort to empower people.[2]

So it should have been no surprise that when the Afghanistan withdrawal quickly turned from a military operation into a dumpster fire, I turned to Jim Hake. No surprise that his team was already on the ground in central Asia, scrambling to figure out what could be done. Also no surprise, Jim told me that if you needed to get things done, "Talk to Chandler."

If you could bottle Terrell Chandler, you could make a mint (maybe not Elon Musk–level rich but certainly enough to retire comfortably to Boca). She is a six-pack of operational expertise. She is also a great exemplar of a networking warrior, though she does not have much of a social media presence (mostly because she is too busy being a real-world presence). She is, however, first and foremost the kind of person you want to emulate if you want to win networking wars.

From Networking to Working

In the film *Forrest Gump* (1994), the unlikely simpleton hero winds up living an incredible, eventful, and impactful life. At one point, Forrest takes up running and manages to attract an army of followers jogging with him across America. Here is your challenge: Be Forrest Gump. Get people to follow your lead. More important, be more than Forrest Gump. He had no particular purpose in mind. Be better. Get people to get things done—for a purpose.

There is zero utility in networking for national security if the network isn't used to do something. It is impossible to dominate without deeds. If the network is just connecting with a lot of people for their entertainment, gratification, sharing casserole recipes, artistic expression, or battling mobs in Minecraft, that is not purpose-built networking. What makes a network purpose-built is having the ways and means to generate action—with a specific objective to impact the outcome of a competition.

Getting a network to have impact requires operationalizing the network. The term "operationalize" means simply to make something operational, conducting a series of activities that lead to an outcome. Back to the military analogies because operations are the bread and butter of military action—that is how the people under arms gets things done.

An operation starts with a plan. Ask George C. Marshall. Perhaps one of the most brilliant, talented, and dedicated leaders ever to serve in the US military, Marshall knew a thing or two about operations. He served in combat during World War I and was the officer responsible for planning the largest military operation executed by the US Expeditionary Force (the Meuse-Argonne Offensive, September–November 2018). In World War II, he directed the army staff. They called him the "architect of victory." How is that as a résumé for an operations planner?

After World War I, Marshall was assigned to the US Infantry School at Fort Benning, Georgia. While there, he kept a little black book, jotting down the names of promising officers, whom he would later select as some of the top military commanders in World War II. Marshall also directed the writing of a book on the lessons learned from fighting in World War I. One chapter was called simply "The Plan." In battle, planning was the "sturdy lifeline of a guiding idea" from this "spins an intricate web that binds an army into an invincible unit embodying a single thought and a single goal." I know, I know. I have said some disparaging things about planning like quoting, "No plan survives contact with the enemy" and "Plans are nothing." However, I also quoted Eisenhower's mantra, "Planning is everything." He (a Marshall protégé, by the way) said this for the same reason Marshall insisted on emphasizing the point about planning. In battle, it is easy to succumb to the chaos around when units are "engaged aimlessly." A good plan can help "mitigate such evils."[3] Plans steel you and your team for the struggle against chaos and competitors, facilitating unity of effort, increasing collaboration, and promoting understanding and trust confidence among the team.

Planning is formalized procedures that result in executing a cohesive system of decisions. Planning processes are formal for the same reason that Emeril Lagasse writes down recipes—to make sure nothing gets left out and the results have a uniform character. Establishing a framework to ensure that decisions are explicit and integrated is necessary to produce predictable results. Plans distinguish what tasks must be done, who should do them, in what sequence they should be performed, and what coordination is required. It is just that simple.

There are all kinds of advice out there on how to plan for social media activities. Much of it relates to commercial marketing practices or social activism.[4] These recommendations are of limited utility, not well suited to the appropriate actions related to national security competition. More applicable would be the kinds of organizing efforts that people like Marshall did or my friends in the small central European country took to battle Russian bots or Terrell Chandler direct to save lives in Afghanistan. The reason for that is that they were all designed to drive concerted, purposeful action in a contested environment.

Before you do, plan. When we wrote our first plan for transition team support in 2012, most of the tasks were sketched out by the executive committee operating in conversation mode, sharing emails.

Fundamental elements of the planning process are (1) establishing a planning team, (2) analyzing information and understanding the situation that would impact the formulation and execution of the plan, (3) determining goals and objectives, and then (4) putting a plan together—the who, what, when, where, how, and why of what needs to be done.

How do you plan? There are two kinds of planning—deliberate and crisis planning. Both have utility for social networking warriors.

DELIBERATE ACTION

Deliberate planning is appropriate when there is time, information, and the other resources required to analyze and make decisions in a formal, intentional manner. Most of our planning for the transition teams was deliberate planning for a simple reason—in each case, we started well before national elections. We had plenty of time.

The process of deliberate planning is straightforward. There are three parts to the planning process.

> (1) Analyze the task. Understand all stated and implied subtasks required, as well as all internal and external factors that impact, help, or stymie getting things done. Take our transition support teams as an example. Some of the tasks we knew had to be accomplished were self-evident—like we needed policies to promote. Other tasks were "implied" by the duties we knew needed to get done. For instance, in 2016, we fretted that the State Department was incapable of implementing the policies we were proposing, which prompted asking Brett Schaefer (discussed in previous chapter) to write a paper on how to reorganize the department to do what we thought would be needed to be done. In addition, we had to evaluate all the factors that might impinge on

executing the plan. For example, did we have the budget to do this? Well, since we did virtually all of our coordination online using existing platforms, our costs were near zero. So we were in good shape budget-wise. That was good since we had no money to budget.

(2) Develop options. Think through alternative, distinct, and feasible courses of actions that might accomplish the task. Evaluate and compare them to determine their advantages, disadvantages, and relative merit in accomplishing the task. This part is as much art as science. Remember back to chapter 2 which discussed the tool kit of tactics—offense and defense, direct and indirect approaches, centers of gravity, culminating points, and simplicity. They are like an artist's palate, there to make choices for developing options. There are always different ways to solve a problem. The art of planning is to figure out the best one. In the case of transition planning, once we figured this out in 2012, we pretty much just followed the same course of action in the following presidential cycles (though we refined the plan each time based on lessons learned from the previous election).

(3) Decide. Select a course of action and then write a plan to implement it. In 2012, for instance, the executive committee drafted a "terms of reference" (the basic who, what, when, where, why, and how) to guide network activities.

CRISIS DECISIONS

This planning process is for when there is limited time and information available, yet there is an urgent imperative to act. Acting in a crisis consists of several crucial steps. The first is recognizing there is a "crisis" to begin with and there isn't time to go through a deliberate planning process. This is not a trivial step. If this decision is screwed up, likely nothing will go right after. The Afghan withdrawal, for insistence, was over in days. If we treated the unfolding disaster in a more deliberative manner, took time to organize a committee, grabbed a latte, and contemplated our options on a whiteboard, the Taliban would have been sweeping out the airport lounge before we had a plan.

Once you know you are in crisis mode, the next thing to do is act that way. That doesn't mean panic or react impulsively. The second action in a crisis is getting your game face on, being serious, focused, and decisive—selecting and implementing a course of action. Unlike deliberate decision-making, in a crisis there is limited time for analysis, study, and consideration of alternative options. So, consider and pick a way to address the crisis. Here, having experience helps. Experience and judgment can be paired to select the best option intuitively and more quickly. Then give the network the "who, what, when, where, why, and how" that you can and get to it. Our Afghan crisis group focused on one task from the start—sharing information. We picked that course for the simple reason that it was the immediate and practical action we could take that might help make a difference.

Finally, in crisis mode it is crucial to communicate explicit, clear, and timely actions. Failing to communicate effectively and rapidly can undermine the legitimacy of decisions, slow action, and put people at risk. In our Afghan crisis group, I took that task seriously. If critical information didn't get passed on or the right folks didn't get connected, that might cost an opportunity to save a life.

Planning Pitfalls

Planning sounds easy. There are, however, so many ways to screw up planning that it is not funny. There are lots of "shelfware" plans out there—plans that are put on the shelf and forgotten. Some don't pass the laugh test. They lack all the key attributes of a successful plan (suitable, feasible, and acceptable), or they don't address all the key questions—who, when, where, what, why, and how. Knowing how important it is to get planning right, how does that happen? It happens because we are all human, and planning is very much a human activity, where the abilities of the planners really matter. Some of us are better at this than others. Some are not trained or skilled. Some are incompetent, corrupt, or stupid. Some, like George Marshall and Terrell Chandler, seem born to do this. For sure, one thing I do know is what it looks like when things go wrong (conversely, doing the opposite usually leads to satisfactory results). Here is the all-star short list of what to avoid from the bitter experiences and occasional successes of years of planning scars and nightmares.

Leaders not engaged in the planning process. This is such a planning buzzkill. All too often, leaders become engrossed in day-to-day activities and either neglect planning activities or delegate everything to subordinates. Shocker, but plans that lack the guidance and commitment of leaders are more difficult to implement.

Misdiagnosing deliberate and crisis planning. Plans are useless if they do not conform to the conditions "on the ground" (whether that is dirt or digital space). Don't be deliberate in a crisis. Don't wing it when you should be deliberate. Some problems are an admixture of deliberate and crisis conditions. Sometimes, conditions flip-flop between one set of conditions and another. There is always tension between having disciplined, routinized, structured planning and the need to modify, improvise, and adjust to meet changing situations. If you don't know what kind of planning you should be doing when, you are more than likely to fail.

Absence of structure and focus. An unstructured, undisciplined planning process without a set timetable and deliverables is like a meeting with no agenda, no end time, and plenty of doughnuts. If anything worthwhile happens, it will be by accident.

Lack of stakeholder involvement. It is shocking how people make plans for other people and don't include those people in the planning only to find out later that if they had, they might well have come up with a better plan. Not to mention, people tend to have little faith in a plan when they played no role in putting it together. Leaving out the people who should be part of the plan is bad planning.

Not evaluating planning adequacy. Plans at the front end ought to think about "feedback," ways to determine if the plan is working as planned. More on this in the next chapter.

Finally, having a plan doesn't mean the plan will get executed and, if it does, work. The Germans started World War I with a brilliant plan. It was called the Schlieffen Plan (August 1914). It would enable Germany to win a two-front war. Except, it didn't. Instead, Germany bogged down in years of attrition combat and lost. In contrast, the Allies in World War II had a sophisticated plan for storming Omaha Beach in Normandy. Almost everything went wrong. They won anyway. Let's cover what accounts for the different outcomes. They are important and as relevant to networking warfare as real war.

Role of the Leader

It is hard to oversell the importance of leadership in turning a plan into action and operationalizing a network. In any network of any size, leaders matter.

What makes for a Terrell Chandler? What makes for a leader that the network can depend on to deliver purpose-driven action? It takes someone exactly like her, not her. What that means is that you need a leader possessing the core attributes of leadership—who can operate effectively in the particular operational environment where things need to get done.

The right kind of leader for any situation, any task, in any network is the right kind of leader. If leadership was a cookbook, everybody would be making leadership cupcakes. We all know from everyday life that is not happening. Everyone knows at least one person who was put in charge of something and delivered leadership flatter than a drive across Kansas. Indeed, faulty leadership often seems more the norm than the exception. After all, somebody came up with New Coke (1985) and people followed their lead. And look how that turned out.

Leadership, as noted before, is a relationship between the leader and the follower. What works best could be different in different situations.

Returning to the importance of skills, knowledge, and attributes of a national security networker—What are the ones needed from the leader? There are three.

Attribute. Not surprisingly, the most vital attribute for a top national security leader is character. For starters, when looking at leadership, it is important not to conflate characteristics and character. They are two very different aspects of leadership. Characteristics are the superficial qualities of leaders themselves. The study of General Douglas MacArthur, one of the most famous American commanders in World War II, offers a case in point. MacArthur had a genius for war. Genius, however, is not a synonym for perfection. Error-free decision-making is not the measure of great leadership. Mistakes in war—even the most awful decisions that, in retrospect, seem maddeningly obvious—do not necessarily reflect a lack of brains or character. Missteps in war are inevitable. And the more responsibility, the longer the campaign, the longer the career, the greater the stakes, and the more far-reaching the objectives, the greater the likelihood that more misjudgments will be made. No great leader has a perfect batting average—MacArthur, least of all. From that frustrating fact arise numerous disagreements among military historians. What complicates understanding MacArthur was his complicated personality. Douglas MacArthur was vainglorious and mercurial. A borderline narcissist, he could be vindictive and petty. Some found his characteristics admirable. Others found them creepy. Many studies of the general became so consumed with explaining, condemning, or excusing his character that it clouded their judgment in evaluating the character of his underlying leadership, the decisions he made, and the impact they had. These writers missed what matters. Characteristics matter way less than character. Character goes to the deeper aspects of judgment and values.

Good strategic leaders are people of character—leaders who do things for the right reasons. Of course, the right reason depends on whose side you are on. If you are a bad guy, it might mean destroying democracy in Moldova. If you are a good guy, it should mean keeping the United States free, safe, and prosperous.

As discussed in chapter 1, national security problems are complicated, that gives rise to a variety of probable solutions, each with their own rationale.

Leaders of character put priorities in perspective and keep them in order in deciding what to do. While they recognize that multiple goals must be addressed, they make sure that lesser concerns don't compromise higher values. In the national security topology laid out in *Digital Dominance*, the top priority is always what is in the best for the nation—not the administration, any agency, political party, stakeholder, or other constituency. As a second priority, leaders must decide what is in the best interest of their organization or network overall. After all, if the group isn't powerful and successful, it is hard to get anything done. Third, leaders must account for what is best for individual members or supporting constituents. People who support the leader are most motivated when they feel they are valued and what they do is valuable. Accounting for their interests is important (see discussion on value proposition in chapter 4). Great leaders put themselves last. They are selfless servants.

Leaders often go wrong by sliding into a twisted pretzel of logic that shifts and inverts priorities to fit what they want rather than what the nation needs. For example, thinking that "what is good for the White House is always what is good for the country" is a case of inverted thinking. Similarly, a team member's drive to succeed in their role should not get in the way of other, more important network priorities. It takes the courage of character to stay on course.

Take the example of Harry Truman who, as president (1945–53), had to make some of the toughest foreign policy calls of the modern era, decisions that short-circuited his chances for another presidential term. After deciding to fight back in Korea, Truman committed a series of blunders and miscalculations—from the decision not to pursue a declaration of war to mismanaging MacArthur. His popularity plummeted. Despite the calamities, writes historian Robert Ferrell in his 1996 biography of the president, *Harry S. Truman: A Life (Give 'em Hell Harry)*, "it was impossible to redeem the errors of the past, but it was possible to make the right decisions for the future." Truman steered a steady course between avoiding widening the conflict and not abandoning South Korea, leaving the next president a salvageable situation. Throughout his trials, with great personal effort and integrity, Truman managed to keep the needs of the nation first in his priorities. Consequently, he made the most of the final years of his tumultuous presidency.

Knowledge. For decisive, impactful leadership, wisdom matters. Leaders need to be proficient in their subject area. They don't need to know everything, but they must know what they do know, what they don't know, and how to cover the gaps by gaining credible, relevant information and assimilating it effectively. The best national security professionals are leaders who "know themselves," understand their own strengths and weaknesses and know how to best fill the space in between.

There are few better examples of high-level proficiency in national security thinking than the launch of the Strategic Defense Initiative (1983). Missile defense was one area where the president, Ronald Reagan, second-guessed his own policy team. He educated himself enough on the issue to recognize the potential to introduce a game changer into the strategic competition with the Soviet Union. Then, writes scholar Paul Lettow in *Ronald Reagan and His Quest to Abolish Nuclear Weapons* (2005), the president began to dig deeper, "acquiring support for the general idea of a missile defense effort from elements of the bureaucracy whose backing and technological assessment he thought he needed to proceed, particularly the NSC [National Security Council] staff and the Joint Chiefs of Staff." Reagan's effort was a perfect example of blending knowledge and learning to achieve an extraordinary policy accomplishment.

Skill. No leader skill is more important than sound professional judgment based on the ability to think critically. A résumé, a list of education and jobs, alone is no predictor of success. Effective national security leaders rise above their parochial backgrounds and personal preferences, demonstrating the capacity for critical thinking at a higher level. Not all leaders can do that. Critical thinking becomes sound judgment when leaders show the capacity to make credible assessments based on available evidence and fuse that with the ability to make timely decisions and follow through on them to produce right actions. The best ones do it under stress and never lose their Steve McQueen cool.

Here is an example of getting it wrong. In a study on the run-up to the North Atlantic Treaty Organization's intervention in Bosnia (1995), researcher Vicki J. Rast details how President Bill Clinton's administration suffered from paralysis by analysis.[5] The failed 1993 US intervention in Somalia haunted the president's team. With the "Somalia syndrome" weighing heavily on their minds, the White House struggled with the decision to commit forces to another intervention with an uncertain outcome. They "overthought" and dithered while the crisis spun out of control, and what were once viable options were overcome by events. In the end, events forced the president to act. Critical thinking is not an excuse for inaction—it is a prelude to correct, decisive action.

The skills, knowledge, and attributes listed here may not be sufficient and comprehensive enough for your network. What you need depends on the mission and what you expect the network to do. When it comes to national security, however, our list here (character, competence, and critical thinking) is as good a place to start as any.

How to Lead a Network

What leaders do is important. What everyone else in a purpose-built network does is important as well. So, one critical question that must be addressed in operationalizing a plan is what kind of connectivity do you need between the leader and the led? The answer is, it depends.

Organize. There are two basic ways to operate a purposeful network—centralized or decentralized. Centralized means the authority for decision-making is concentrated in the leader, who then basically just tells everyone else what to do. Decentralized is the opposite. There is very little specific direction from the top. Folks mostly figure out for themselves how to get things done.

A few factors help determine which mode works best. Mostly, the decision to centralize or decentralize is based on risk. Competence and consequence matter a lot. On the one hand, if networkers are highly skilled, it is easier to delegate with confidence. On the other hand, if the consequences of the success or failure of a task is critical and the network is less skilled, central direction might be more in order. On the other, other hand, if the consequence is low and the abilities of networkers are low, decentralization might make sense anyway. Let folks learn from their mistakes; that will either help build better leaders for the future or weed out the weak ones.

Master Intent. One of the great advantages of networks is that they can facilitate joint action, collaboration, innovation, speedy adoption, and adaptation. Our Afghan crisis network was a good example. It was about as decentralized as you could get—one step from chaos. There are inherent advantages to operating in a more distributed fashion. One tool that facilitates decentralized action is effectively and clearly sharing the leader's intent behind the plan—understanding why a task is important. If competent members of the team understand why and what the leader wants to accomplish, they might on their own figure out a better way to accomplish the goal. The Schlieffen Plan failed in 1914 because German commanders just did "what" they were told—and that didn't work. When the Germans next invaded France in 1940, they followed a practice called *Auftragstaktik*, where commanders were expected to understand the purpose of the mission and adapt their actions and the plan to fit the conditions they confronted. So instead of following the plan exactly as written, they pushed their main effort through a weak spot in the Ardennes Forest and pretty much rolled into Paris.

There is a rich literature on "leader's intent" in the military, business, and leadership literature.[6] It is worth consulting because the more you can "power down"—in other words, the more you can decentralize execution in your network—the more responsive, innovative, and adaptive your team will be.

Prepare. There is a tendency in social media to just have at it and learn by doing. You can do that. In fact, there is something to the adage that experience is the best teacher. Still, the odds of success increase if you know what you are doing.

For starters, you can train leaders and networkers in the skills, knowledge, and attributes they will need to be successful. One of the first steps I took when we decided in 2021 to run a transition support network for the long run—right up to the establishment of a formal project for the start of planning for 2024 elections—was to organize our networks into committees, appoint a head of each committee, and school them on the elements of building a network that are summarized in several chapters here: recruit, value proposition, operationalize, and feedback (covered in the next chapter).

Stress Test. There are different ways to test your plan and determine if it will work or how to make it better. A plan can be practiced or "war-gamed." A war game is free-play simulation that allows for assessing the outcome of actions in a competitive environment. There is a rich literature on the topic with applications for everything from fighting real wars to responding to a cyber breech.[7] Operations can also be "red teamed." A red team exercise employs a group of simulated adversaries or experts to look at your plan and determine its weaknesses and how to disrupt or defeat it. The plan can then be revised and made stronger. There is vast how-to literature on this practice as well.[8]

Learning. Another way to improve operation performance is lesson learning—reviewing the operations already performed, seeing how successful they were, and figuring out how to improve on them. After we got our transition network organized into committees, I had each of the heads of the committees share their experiences with one another so they could learn from one another. More on that activity in the next chapter.

Know the Rules

One bit of advice is to follow the rules. Chapter 2 made the point that true national security competitions aren't bound by rules. Networks on social media (except, of course, the covert and criminal ones that ignore the rules), however, are bound by the laws of the places they operate in. Not following them risks the credibility, integrity, and perhaps even the legality of the networks.

If staying legal is important to you, pay attention to this part. Before going operational, know the rules. All social media platforms, for instance, have terms of use that define permissible activities. There are also a plethora of regulations, laws, and requirements that might impact network operations. For example, lobbying activities in the United States on behalf of foreign governments are managed under the Foreign Agents Registration Act.

Let's go back to the example of our transition support network. It was a completely informal, volunteer group, so there were not many rules to worry about. On the other hand, the organizations that supported our work, like providing platforms to network on, did have rules to follow. As 501(c)(3) organizations (under the US Internal Revenue Code), we were all operating as nonprofit, nonpartisan entities that were prohibited from doing certain activities, including political action. So it was important that we steered clear of any campaign activities. We focused on the transition, which in contrast to what was going on along the campaign trail was not considered a political activity under the law. Whereas following the rules wasn't a big challenge for us, for other networks it might be much more complicated, particularly if they are operating in the law enforcement, defense, or intelligence spaces.

How to Operate a Network

Follow the plan. When the plan doesn't work, figure out why and adapt. The steps described above must seem formulaic and linear-first: Do this, then do that. Don't think of them that way. Structure and process don't have to be synonymous with rigidity and lack of imagination. In the same vein, innovation, adaptation, spontaneity, and experimentation don't have to mean anarchy. The genius of dominant networking is to merge the advantages of linear and nonlinear approaches together and exploit the best of both. Good planning, leaders, and operations are there to facilitate action, not weigh down the team.

Here is an example. After 9/11, Congress established the Department of Homeland Security. After observing the department once it got up and running, it seemed clear that some of the decisions Congress had made in the enabling legislation on how the department ought to be organized were dysfunctional. Our think tank crafted a plan to address that. We partnered with another think tank, so it was abundantly clear that our effort was bipartisan. We pulled together teams of experts. We gave them detailed terms of reference on what we needed to be analyzed. They reported back with recommendations. We wrote a comprehensive report on how to reorganize the department.[9] We testified before Congress on the results and then started advocating for our findings.[10] Our expectation was to get Congress to go back and amend the authorizing legislation. Then something unexpected happened.

The secretary of homeland security left, and the president appointed a new one, who announced his priority was reorganizing the department. He did. Many of the recommendations we made were implemented. How do we know we had an impact? He told me so. The first time I met the new secretary, he shook my hand and said, "I know who you are. I read your report. When I was nominated, they gave me a bunch of stuff to read. I read your report first because it was short." He was all in. We had not anticipated that was how we were going to get results. But we adapted. In the end, about 80 percent of what we called for was put into action. The point was, we didn't just write a report. We took an idea. We had a plan. We engaged in purposeful action. We rolled with the punches. That's how real operations work.

How to Protect a Network

The idea of protecting a network was first introduced in chapter 3. That chapter presented the concept of risk. At this step in setting up a network, networkers need to pay attention to the risks they may face.

Depending on how serious your enemies are, operating a network could incur significant danger. Once a network does anything, its activities will draw attention that can, in turn, attract attacks against the network. Future chapters will add a lot more on attacking and defending networks, but it is worth laying out the fundamentals here. The most basic guidance of all is protect first and foremost what allows the network to operate. How much protection is needed is dependent on how much risk the network can tolerate (for a discussion of risk, see chapter 3). At the very least, take the commonsense protection efforts that people ought to take in their everyday activities.

The most basic of these is cybersecurity. Whether networks are operating online or offline, odds are they are using some digital devices with access to the internet. It might seem like common sense that most people would take commonsense precautions. They don't. Indeed, younger generations who have grown up with social media and digital technologies are the most careless. They fall for scams at higher rates than senior citizens.[11]

The basics of cybersecurity, which are basic to any network operations, are called "cyber hygiene," rudimentary security habits and practices.[12] There are different lists, but a representative one would include the following.

Network Defenses. This includes a network firewall (prevents unauthorized access), data-wiping software (clears out unneeded system files and data), a password manager (keeps all the users straight on security rules to follow), and antivirus software (detects and removes malicious software).

Password Practices. Avoiding the stupid mistakes that allow passwords to be easily guessed or compromised. This includes important things like not having a password called "password" or "12345." Here, by the way, is the most common list of passwords used in 2021.[13] If you don't think checking this list is the first step that every cybercriminal will take, you really need help.

Multifactor authentication. Requiring multiple security features to enter a network. An example of this is selecting the feature that requires entering a text code sent to your cell phone before an app will let you sign in after you have entered your password. A little extra time-consuming effort in signing in adds a lot of protection.

Data Backup. Secure and protect files against data loss by backing them up to an external hard drive or cloud services. Nothing is worse than having to do the same work twice because it gets lost in the ethernet or, worse, stollen by your enemies and used against you.

Update Systems. Ensure that software updates, which often include new security features, are installed. If software and service providers are giving you this stuff, usually for free, you are an idiot if you don't use it.

Secure Routers. Don't use unsecured Wi-Fi.

Avoid Social Engineering Attacks. Don't be stupid. Here is one recent list of the most common types of attacks.[14]

This list is not meant to be comprehensive, but if these recommendations don't strike the reader as commonsense practices, you need to get yourself a list. This advice, by the way, applies not just to cybersecurity but to all the essential functions of a network (like physical security) where commonsense protection makes sense. Do the basics no matter how unobtrusive, humble, or harmless you think your network might be.

Mental Health. Social networks are about humans. If humans can't operate with efficiency, that degrades the network. Ensuring the physical and mental health of networkers, particularly in high-stress activities, might not be an insignificant challenge. Post-traumatic stress disorder (PTSD), for example, comes from "stress," and stress can be generated by a variety of both physical and mental activities either online or offline. Remotely piloted vehicle (drone) operators, for instance, experience PTSD.[15] During COVID-19, researchers found that the rapid expansion of online activity in schoolwork contributed to stress, depression, and anxiety.[16] Stress is everywhere. Mental health risks that impact network performance should not be ignored.

Serious mental health issues of course require professional medical care. Just like with cybersecurity, however, there are basic mental hygiene activities that will protect the networking forces. There are a lot of lists. Here is a short one.

> Eat healthy and exercise. The link between mental and physical is undeniable.[17] Big yoga fan here.[18]
>
> Rest. The human body requires a minimum of four hours of uninterrupted sleep to continue to function. That's the minimum even during the most severe crisis. In general, a consistent schedule of sleep, seven to nine hours each day, has significant benefits from reduced stress levels to a healthier body.[19]
>
> Socialize. Interacting with other humans outside the networking and work environment, as well as engaging in other activities including relaxation, music, and meditation help reduce stress.[20] Take five and socialize.
>
> Get Outside. A twenty-minute walk in the park is better than a valium. Okay, I made that up. That said, there is tons of research that makes the case that being outside in nature relieves stress.[21]

Taking care of people and systems will always be fundamental to protecting networks and keeping networkers in the fight.

Offense and Defense

This is the point in social networking where the concepts of offense and defense have most relevance. If a network is ever going to drive action and change, it is going to have to operationalize, to go on the offense and do something meaningful. In contrast, in a competitive environment, as soon as you start operating, others are likely to notice, and if you are having an impact they don't like, they might well act against you. You may have to defend yourself.

Social networking for national security is like war, a real competition. In future chapters, there will be a lot more detail on how to attack networks and conversely how to defend against attacks. Not surprisingly, the focus will primarily be on operational activities. They are the most observable and important actions in a network. What will likely be the focus of action, either in the context of offensive measures or adversarial action against a network, will be the key elements of operationalizing a network that have been discussed in this chapter—planning, leaders, operational activities, and protective measures. Getting these wrong or right could make all the difference. Let's look at some military examples as an analogy.

Planning. One of the greatest Allied victories of World War II was the Battle of Midway. One big reason why the US Navy kicked butt? Cryptologists decoded the Japanese secret code and then through a clever ruse, uncovered the enemy plans by determining where the Japanese were going to attack. After compromising the Japanese scheme, the navy came up with a kick-ass plan all its own, mounting a counterstrike to surprise and defeat the Japanese fleet.

Leaders. The Japanese military genius who organized the attack on Pearl Harbor and almost bested the Americans at Midway (if he only hadn't been beat by a better plan) was Admiral Isoroku Yamamoto. In 1943, US fighters shot down the plane ferrying him to a meeting, thus robbing the enemy of their greatest naval leader.

Operations. With Yamamoto out of the way, General MacArthur undertook his "island hopping" campaign, bypassing Japan's main defense positions in a series of operational actions that gave the Americans a decisive advantage the Japanese could not match. MacArthur leapfrogged all the way to retaking the Philippines.

Protection. After Yamamoto met his maker, MacArthur ran wild throughout the Southwest Pacific because by that point in the war, the Allies had beaten the Japanese air and naval forces so badly that the United States could pretty much protect its troops from attack. When one side is protected from the other, making war is much easier.

In short, going after the other guy, safeguarding plans, leaders, and operations allows for taking the initiative. Conversely, undermining the enemy's capacities hamstrings an enemy's ability to act. Future chapters will dig more into how to do that as part of network warfare.

Know at this point in organizing network activities that if you are operating in contested space, you will likely have a fight on your hands. The next chapter will explain how to better ensure your network can compete, survive, and thrive.

6

FAILURE TO COMMUNICATE

Bob Goldfarb is a networking ninja.

That is some achievement since he was fighting America's wars almost half a century before most social networking apps appeared on the internet. He was way ahead of his time.

Of all the mentors I have had in my life, Bob taught me one of the most valuable life lessons. When it comes to dominating action, he schooled me in understanding the most neglected part of standing up a resilient and impactful team.

History made Bob Goldfarb an extraordinary person.

What do you do after you fight a war and aren't killed in the process? Robert W. Goldfarb did what most of the Greatest Generation did. He went home. He raised a family. He got a job. He went to work. That was where Bob learned that he had learned the secret to great work—when he was back in the military.

During the Korean War, Bob served in the ground forces. The strength for their fight came from the ability of small units to work together, improvise, adapt, and persevere under punishing conditions. Like a lot of veterans, Bob brought those lessons learned into the workplace.[1] The most important ones had to do with understanding the relationship between the leader and the led. "Other than marriage and fatherhood, nothing shaped my character more than my three years in the army," Bob recalled. "The men who trained me had jumped into Normandy. When one of them told me 'You are one of us,' it felt like a benediction imposing an oath that I would not disappoint him, then or ever."[2] In both his training and wartime experiences, Bob learned the value of leaders and teams.

Goldfarb's particular passion was coaching leaders to lead successful organizations. He founded Urban Directions, Inc., a management consulting firm that he ran for over thirty years. He wrote a book about what he learned in working with the CEOs and managers of some of the top companies in the world, *What's Stopping Me from Getting Ahead?* (2010).

Even though he is long retired, Bob has not stopped mentoring. He works with military veterans to help them find successful private sector careers. Bob guides them through closing the gap between the preconceptions of civilian employers and the frustrations of combat veterans. "I could not stand idly by," Bob shared in an article he wrote, "while those who might have worn the same shoulder patch I did were being seen as potential employment risks simply because they had served." He found that "over 40 percent of them [veterans] leave their first job within a year, frequently citing the difficulty of adjusting to a culture very different from one they are accustomed to. . . . I am devoting myself to helping them prove they can be as competent in the workplace as they were on the battlefield."[3] Last time I checked, Bob is still out there and at it.

I know Bob because he was hired to consult for the think tank where I work. He would periodically come in and coach individual members of the management team—ask what our challenges were and how he could guide us in helping ourselves. Bob also asked how we viewed the performance of the organization overall, what we saw as the obstacles and opportunities. Okay, here was the deal. Our conversations with Bob were confidential. That said, after Bob met with us, he would report back to the CEO. We knew that. We knew Bob knew that we knew that. The CEO knew we knew, too. That was okay with the CEO. That was exactly what he wanted. The CEO valued having a nonconfrontational, honest, anonymous channel of feedback, a safe way for managers to share what they were really thinking. Of course, we had management meetings and sent in reports, but this informal means of communication served as another avenue for the CEO to take a pulse of how things were going. I always suspected that this was Bob's idea, but I know the boss found this helpful. After he retired, he told me so.

These practices and Bob's feedback must have worked. Our think tank was featured in a book about successful leadership in the nonprofit sector, Leslie R. Crutchfield and Heather McLeod Grant's *Forces for Good: The Six Practices of High-Impact Nonprofits* (2007).

In our many conversations, when we talked about my own management challenges, Bob would often ask me to reflect on my military experiences. "How would Captain Carafano have dealt with this?" Then he would ask, "How does this relate to what needs to be done here?" I always thought he brought up the military because it was something we had in common. It was, however, more fundamental than that. Bob was keying on the things that successful organizations, any kind of successful organizations, could do to be successful—whether on battlefields or in boardrooms.

Often, our discussions came back to the issue of feedback—whether it was counseling folks on their performance and how to improve, or getting more teamwork in the team, or better communications between leaders. All the coaching focused on how to help the team get things done and get them done better.

Feedback Loops

When it comes to feedback, the movie that comes to mind is *Cool Hand Luke* (1967). A sadistic prison foreman brutalizes Paul Neuman while repeating the famous cinematic refrain, "What we have here is a failure to communicate." Can't ask for more direct feedback than that.

Feedback is particularly crucial to systems for improving performance. A feedback loop simply means a way of measuring outputs that provides insights used to make decisions and guide future inputs. For example, if you buy stock online and then check to see if the price per share went up or down and then based on that feedback decide to buy more stock or sell, that is a feedback loop.

Feedback loops are employed in all kinds of systems, linear, nonlinear, mechanical, and human. What they share is harnessing an understanding of performance to affect future outcomes. For example, one study of Canadian fisheries examined how feedback could be used to improve the management of wildlife resources.[4] Whether it is counting cod or measuring members in a network, the basic concept is the same.

Military units are a kind of system. Inputs are orders. Outputs are following orders. When Bob Goldfarb was in the army, before his team headed out on a mission, the sergeant would conduct a precombat inspection. Having ordered the men to get ready, he inspected their weapons and equipment to see if they had done what they were told to do to the standard the sergeant had trained them. That was feedback for the sergeant—helping him know how well his orders had been followed. If the orders had not been followed correctly, the sergeant would adjust the inputs. That might take the form of a hard slap on the back of the helmet and few unrepeatable words warning not to screw that up again. That was feedback for the soldier. I always imagined that was where Bob learned a thing or two about feedback.

Networks are also a kind of system. Feedback loops can affect their performance. Feedback loops are especially important for networks dealing in national security. National security networks are purpose-built—organized for one reason—to deliver outcomes. Feedback is particularly import to these kinds of networks because delivering on the outcomes is an important contribution to ensuring national security. No one, for instance, has much interest in a missile defense system that can't shoot down missiles. Outcomes matter.

The advantages of a feedback network for networks are manifold. For starters, feedback helps address the slacktivist problem discussed in previous chapters. If feedback loops are structured properly, they can provide visibility on whether participants in the network are doing what the network wants. Indeed in some cases, having a feedback loop helps promote action because users in the network know there is an expectation of action and people in the network will know that others will know whether they followed through or not. Here is an example. A previous chapter offered the case of the insurgent network that attacked American troops in Iraq with improvised explosive devices (IEDs). One link in the network was a videographer who was stationed at the scene of the attack to film the explosion. That video served a couple of purposes. For one, the bombers didn't get paid unless they could prove they attacked someone. That was a form of a feedback loop. Feedback also helps improve performance. The videos of the improvised explosive attacks were studied to assess how to improve future attacks as well as evaluate the state of countermeasures to elude their attacks.

Feedback can also be a network asset. In the case of the improvised explosive bombings, the videos were posted online to show how powerful the insurgents were and the havoc that they were heaping on the Americans. They proved to be of immense propaganda value. In addition, they powerfully reinforced the insurgent network's value proposition (discussed in chapter 4). Remember, a good value proposition is clear, explicit, and backed up with evidence. The insurgents' value proposition was, we will enable you to kill and maim Americans and get paid for it. What better proof of that than an actual video showing an American armored vehicle disappearing in a ball of fire, enveloped by a mushroom cloud of desert sand? In short, even the measure of outcomes can be a useful outcome.

Measuring

As was discussed in previous chapters, measuring outcomes in a productive manner for a national security network can be a challenge. Does the activity of a network lead to action? That is the big question.

Measuring outcomes in social networks can be tricky. Many of the public feedback elements on social media platforms are pretty much useless for national security purposes. Likes, views, clicks, open rates, comments, ratings, and so forth might give some indication of how many people are paying attention to what is being said, but that information isn't all that helpful. Useful feedback must measure what people do, specifically measuring activities relative to the accomplishment of the network's mission.

Performance metrics (setting measurable goals that can be quantified) may also not be suitable as a means for providing useful feedback. Performance metrics were all the craze back in the day, taught in every business school as the Holy Grail of management. It was like putting leading on autopilot. Set a performance requirement, like how many cars a dealership must sell. If the dealership hits their quota, all was good. If they missed, fire the salesperson. If they exceeded sales goals, it was Christmas bonuses for everyone. The US government became so enamored with this magic elixir of management that in the 1990s, the administration mandated the use of performance metrics throughout all federal agencies. Here was the problem with that.

Performance measures were useful for measuring linear activities, like how many social security checks were issued. They weren't terribly useful for nonlinear activities like "be more innovative." Even in just measuring linear outcomes, performance metrics might be unhelpful or even counterproductive if they measured the wrong things. Sometimes there is an impulse to measure what can easily be measured, rather than measure what really matters. Furthermore, some managers are happy to measure things when they know they can meet or exceed the performance objectives. They get less excited about measuring things that are difficult to do—even if they are more important. Delinked from focusing on the mission, performance metrics often devolve into measuring outputs (like how many posts on a platform) rather than outcomes (like how many people acted on the advice posted). For national security networks, there must be a better way.[5] There is.

There are proven techniques for establishing constructive feedback loops for all kinds of network activity. Some of these were mentioned in the last chapter on operationalizing the network. Others draw from a variety of fields where network feedback is crucial to mission success.

Operations Research. One way to get feedback is to study the system's performance and then give feedback on how to either improve it or, if you don't like the system, screw it up. The operations research discipline originated with Frederick Taylor, popularized in his book *The Principles of Scientific Management* (1909). Taylor studied industrial processes. One of the most famous analysis looked at transporting pig iron at a smelting plant. With a clipboard, pencil, and stopwatch, Taylor recorded each component of the process, measuring each input and output and then determined how to maximize production. At the time, it was groundbreaking research on industrial management that quickly drew a huge following.[6]

Taylor's method of analysis became so dominant that they called it Taylorism. The military adopted this concept for evaluating military systems. The armed forces used what they called "operations research" for everything from determining how to maximize strategic bombing for campaigns like Operation Pointblank or how to disrupt German U-boat operations in the Atlantic.[7] When I was in the army, we had operations research officers assigned to our field commands. The military still uses this technique. Operations research, for example, was the core of the Joint Improvised Explosive Device Defeat Organization (JIEDDO) activities to determine how to defeat Iraqi insurgent IEDs. The method continues to be employed in many civilian applications as well.[8]

Operations research can have some utility in analyzing and providing feedback on networks. How useful operational research techniques may be depends on the nature of the networks. Operations research is most effective looking at linear processes; that is why back in the day when industrial assembly lines dominated the workplace, Frederick Taylor was considered the messiah of management wisdom. The more nonlinear a network is, however, the more difficult it is to measure inputs and outputs, and the harder it is to identify subsystems and how they impact the cause and effect of what the system does. For nonlinear systems, operations research may have less utility in analyzing network activities. Further operations research can be very intensive in terms of the time, resources, and information needed to conduct an analysis. This method might not, for instance, be a practical means of feedback for fast-moving competitions that would be over before the researchers could write up a research plan.

Survey. The simplest way of getting feedback is to just ask. For instance, when we stood up our Afghan response network, I emailed people and asked if what I was sharing was helpful. For sure, surveys can be a lot more sophisticated than that. Surveys are recognized as a key performance indicator tool for operations research. There are many forms of surveys that might prove useful for a network from 360 assessments that measure the impact of leaders by asking for evaluations from their superiors, peers, and subordinates to reading unsolicited reviews on Yelp or asking for feedback using online applications like SurveyMonkey.[9]

Of course, surveys are opinions that may or may not be sufficient or helpful enough on their own. Survey information might have to be combined with other forms of feedback to get a fuller picture of how the network is really performing.

After-Action Review. This a common technique used in the military, but it has been widely adopted by others in business circles and nongovernmental organizations. It works like this. After an action has been conducted, assemble the people responsible for and involved in the activities and answer a series of questions: (1) What was expected or supposed to happen? (2) What happened? (3) Why did things work or not work as expected or worked better than expected? (4) What can be learned from this?[10]

The review process can work for operations in both virtual and real-world environments with a couple of caveats. For one, the individuals conducting the review must be open and honest in their assessments. Otherwise, they aren't really learning; they are promoting or defending their own view of what happened. For another, they must have, between them, adequate information to evaluate cause and effect of what happened. A common error in analysis is to equate correlation with causation. Just because an action was taken (a text was sent) and another action occurred (the Arab Spring sprung) doesn't necessarily mean that one action caused the other. In an analysis like an after-action review to understand the impact of activities and decisions, there has to be enough information to establish a causal relationship between an action and outcome.

Observation. Information may not always be available to definitively link cause and effect. In that case, the best that can be done might be to make inferences and judgments from the information or intelligence that can be gleaned by observing the activities of the network and the surrounding operational environment. Take the example of our friends from the small central European country worried about public demonstrations over planned American military convoys. They gathered the information they could in order to understand the situation on the ground as best they could. They couldn't definitively prove what the Russians were up to, but they took what they knew and made an informed guess of what was going on, and that helped them to determine their response and next steps.

Information may not be readily available to assess network performance. Take the case of our Afghan crisis network. We were operating in real time. Feedback had to be virtually instantaneous to be useful. There wasn't a lot of opportunities for that. We did, in fact, get some feedback and turned the information around fast enough to be useful. At one point, for instance, we got an alert from Spirit of America that there were some unauthorized solicitations of invoice payments on their behalf. This looked like people were trying to scam folks to make money during the chaos of quickly throwing rescue operations together. In response, we quickly issued a warning to others to watch out for fraudulent activity. Frankly, this kind of useful feedback was the exception. Most of the time we were just, as they say, flying by the seat of our pants and hoping what we were doing was having an impact.

In-Progress Review. This is exactly what it sounds like. At some point, before an action needs to be accomplished or completed, a point that still allows for time to fix things if they are not going right, get the people in the network responsible for the action together and check to see how things are going. If there is a developed plan for network activity, as described in the last chapter, this feedback process can be effective because a good plan will detail who needs to accomplish what by when. Review the tasks in the plan, assesses their status, and make changes if adjustments to responsibilities or the plan or the activities need to be made to ensure the outcomes desired can still be accomplished.

The in-progress review was the most valuable feedback tool we had in our presidential transition group. After all, we knew the day of the election. All we had to do was backward plan from there, determining what had to be done when to make sure everything was done before the president-elect waved to cheering crowds as they made their way down Pennsylvania Avenue. We just had to be disciplined enough to periodically pause and check to make sure everything was on track.

Brief Back. The shortest feedback loop is a brief back. Basically, it is asking someone who has been assigned a task, duty, or mission to brief back to you what you just asked them to do. Of course, this doesn't ensure that the task will get done, but at least it is a process to confirm that the tasker and the tasked have the same understanding of what needs to get done.

The brief-back process is particularly useful in crisis situations where there is little time to issue formal written plans or hold interminable meetings to go over everything again and again. In our Afghan crisis response network, we used this tool among ourselves all the time. For instance, I would craft a short email to others saying this is what I think I am doing for you. They would correct me or email back "OK." Then at least as I set off, I knew I was trying to take the actions that were expected of me.

Self-Assessments. If all else fails, ask yourself, "How am I doing?" Self-assessments may not give you all the feedback you need on how the network is functioning, but they can be very useful in understanding your role in the network. A self-assessment is simply a measure of what you are doing and how well you are doing it. For example, if you have established goals or performance measures for your activities in a network, you can assess for yourself if they are being achieved and, if not, why? You can evaluate your decision-making. Are you doing the right things for the right reason? You can assess productivity. Are you making the most productive use of your time and your talents?[11]

Furthermore, self-assessments aren't just for yourself; others in the network can be encouraged or required to conduct self-assessments, share their findings, and plow the results back into improving network performance. For instance, in our national security committees, I constantly badgered the committee chairs to review their recruiting methods and results, pushing them to stay on top of their game, bringing as many individuals as they can with the right skills, knowledge, and attributes who could potentially contribute to the next presidential transition.

Technology. There are all kinds of technology and software that are available that could be harnessed to provide feedback for networks from flying drones over the battlefield to accessing CTV footage at the scene of a riot. In addition, there are many instruments for evaluating data from programs that glean data through open-source intelligence from news reporting to tracking what people are doing on their cell phones. Social media platforms can also be a useful tool—if you know what you are looking for. During the Afghan crisis, for instance, I trolled Twitter (now X) constantly, looking for information that confirmed or contradicted what our group was reporting. Usually, our intel was ahead of the Twitterverse, but on occasion I picked up something that suggested we needed to fine-tune what we were doing.

How to Deliver Feedback

Having a system to gather and deliver feedback is one thing, but delivering feedback in a manner that is useful and constructive—well, that is another thing. There is a rich literature from the human resources world on performance evaluations and feedback that has application here. That makes sense. Remember, the point that *Digital Dominance* has hammered again and again about social networking warfare is that it is not about managing a mechanistic system; it is about impacting human performance and behavior. So it is worth looking at the lessons learned from the human resources world of how to help make humans perform better.

Timely Feedback. Nothing is more useless than feedback that comes too late to impact network performance. Going back to the example of our transition support networks, feedback after the inauguration might have helped us think about how to do better for the next election four years in the future, but it wouldn't have been much help in this go-round. Performance is an ongoing activity. The feedback loop should be adequate to match the pace of operational activities. Bob Goldfarb, for example, usually checked in with us a couple of times a year, strategically timed ahead of critical organizational activities like budget planning, goal development, and personal assessments.

Be Specific. National security networks ought to be focused on outcomes. The more feedback that can be tied to specific outcomes and specific issues and recommendations on how to adjust performance to achieve them, the more useful the feedback is. When Bob was coaching me, he never talked or allowed me to talk in generalities, like saying, "That manager is great or not good." He always drove to nail down specifically what was making our organization perform better or what was holding us back.

Be Holistic. It really helps to understand how specific feedback and the performance of individuals in the network impact the overall performance of the group and the attainment of outcomes. Bob recognized that I was an instinctive networker, constantly looking to build teams across the organization. He kept reminding me repeatedly how this was helpful to others and contributing to the overall success of the team. Others needed to appreciate the importance of networking and how it contributed to mission accomplishment.

Highlight Success. It is normal to inherently look for feedback that identifies shortfalls to fix. Feedback, however, is also important for bringing attention to what works. Research shows that positive reinforcement and recognizing superior performance and achievement incentivizes even better performance. In addition, positive feedback serves as a lesson to others on the behaviors and activities that they should emulate. Bob would often spend as much time asking me, what are others doing well, whom do you admire and why, as he did asking me how I was doing. He was consciously looking for performance achievements to highlight to the CEO, helping the boss understand who the strong horses were pulling the team. Who could the boss point to and say this is the kind of high-level performance we are looking for?

Feedback and Performance

One of the great advantages of social networking across online and offline platforms is the multiplicity of tools available not just to collect feedback but also to deliver that information back to the network in a manner that can improve performance. It is a mistake to think of a network as an entity restricted to a single digital platform. In fact, different networking activities might be suited to different platforms and modes of communication. For instance, individual performance reviews might be better suited to conversation mode, done in person or over video teleconferencing. Other feedback might be more fitting for broadcast mode. For instance, when Spirit of

America alerted us about their unauthorized billing incident, which thankfully they were able to stop before any harm was done, I started sending out generic Twitter alerts warning folks to be wary of potential scams (a frequent problem in humanitarian and civic action activities online, as discussed in chapter 2). What is important is recognizing that useful constructive feedback comes from designing a feedback loop about operational activities in a manner that makes the feedback loop the most useful.

Of course, social networking platforms are not universally and uniformly available. Networks must adjust their activities to the apps at their disposal. One of the advantages of a deliberate planning process, as was discussed in the last chapter, is that it encourages as a preliminary step surveying the capabilities available for the network and factoring that in when planning a course of action. As part of that effort, a good deliberate planner will think through the requirements for feedback and fold that into the plan. When we did our transition planning, for example, we included the requirement to conduct in-progress reviews.

Offense and Defense

Feedback loops are empowering for both offense and defense. Good feedback mechanisms embolden offensive operations. Knowing how things are going facilitates exploiting successes or adjusting and compensating for setbacks. On defense, feedback loops enable making prudent choices to protect against the enemy's advance. One of the key reasons the Union won at Gettysburg was because the Union cavalry, the eyes and ears of the army, spotted the Confederates first and delivered that feedback in an effective manner that allowed the Union commanders to take up commanding defensive positions. In contrast, the Confederate cavalry were nowhere to be seen on the first day of the battle. Confederate commanders lacked any feedback on the action at the front until they heard firing as their troops blundered into contact with the Union cavalry.

The crucial advantage that feedback can play in networking competition was first described in *Wiki at War*.[12] One of the key insights for the book came from the career of the air combat ace and strategist John Boyd. Boyd described a process that became popularly known as the OODA loop. OODA stands for observe, orient, decide, act. The concept originated from Boyd's experience in fighting in the same war as Bob Goldfarb. But whereas Bob battled mud and snow, Boyd sailed far above in a fighter jet, piercing white clouds and blue skies.

During the Korean War, US pilots achieved an incredible "kill" advantage over their counterparts. They shot down ten planes for every American plane the enemy took out. Tom Cruise and *Top Gun* (1986) had nothing on John Boyd. Boyd determined the Americans' decisive advantage was that they could spot, orient, and shoot faster than the enemy. His critical observation was that in a competition, the side that could get critical information first, process the information, and act on the information faster than their opponent usually won, like the faster gunfighter at the OK Corral. This insight has important relevance for social networking. The basic principle is the same. The side that gets the dominating information advantage is likely to come out on top. The crucial link in the OODA chain is feedback: the ability to find, process, and return vital information in a usable format faster than the network that the network is competing against.

Boyd's OODA loop answers the most important question, whether a network is attacking or defending: How fast and how good does feedback need to be? The answer is, better and faster than the network that the network is competing against.

Let's go back to the Iraqi insurgency and the struggle of action and counteraction between the terrorists who were relentlessly trying to blow up Americans and the effort to defeat and counter them. A previous chapter mentioned JIEDDO, the organization stood up by the Pentagon, which directed a herculean data collection effort. They wanted to know everything about everything that had anything to do with IED attacks from the time of day to the extent of damage and injuries. This feedback was collected for one express purpose: to develop countermeasures to defeat or mitigate the attacks better and faster than the insurgents could adapt and field threats against the troops on the ground. The results were a mixed bag. One study by the research institution RAND concluded there were lots of inefficacies in the organization's capacity to absorb all the information they collected and turn data into usable, decisive actions. Others outright claimed the bomb fighters bombed. One conclusion was inescapably true: they failed to use feedback to decisively dominate the competition.[13] In the end, the bombings stopped when the insurgency ended. The example illustrates how well feedback is used can play a significant role in determining the course of a competition. Bob Goldfarb could have told them.

Summary

To summarize, feedback loops are crucial and potentially decisive components of a purpose-built network. There are a number of conceptual and technological tools for providing feedback to networks. Good planning will not only address incorporating feedback into a network but figure out how to deliver that feedback in a manner that is accessible and useful. The key advantage of network feedback is when the network can deliver and execute on feedback faster than the network's competitor.

Now What?

In the last four chapters, *Digital Dominance* has explained how to become the dominating network warrior described in *Wiki at War*. The chapters outlined four components of a potentially effective and dominating network: (1) recruiting, (2) establishing a value proposition, (3) operationalizing the network, and (4) delivering feedback.

We are not done yet. National security is competition. It might not be enough to know how to build a network. To survive and thrive, it might be necessary to take your competitor's network down.

The last four chapters discussed some of the tactics and tools for conducting offensive and defense operations. It is also important to lay out conceptually not only how to use them to build, employ, and protect a network but also how networks can be organized into a campaign to take out the other side's network. Sometimes, winning is about finishing first. Sometimes, coming out on top is about making sure the other side finishes last. The next four chapters go into the components of doing that.

Now we get to the part of the book where we cover breaking things. For Bob Goldfarb, who spent his entire professional career helping people build things better, I am sure in that old warrior's lion heart is still a yearning for the good old days when we were soldiers once and young and a good day was measured by the misery inflicted on your enemies.

PART THREE

How to Beat Networks

7

YOU'RE CONNECTED

Kara Frederick almost got her future husband killed. He married her anyway.

This is what happens when you enter the workforce after 9/11 and wind up in the intelligence business and your role model is Jack Ryan. You end up hunting terrorists for a living. For over six years at the US Naval Special Warfare Command and later for the Department of Defense, Kara developed an expertise as a counterterrorism analyst. She deployed three times to Afghanistan in support of special operations forces tasked with tracking down and eliminating bad guys. These were "hunter-killer" operations. Kara's job was to hunt down terrorists and send others out to kill them.

To find her targets, Kara would sift through intelligence trying to put together a picture of insurgent activities. Once she and her team mapped the network, they would start tracking specific "high value" targets (HVTs), assets that if captured or killed would significantly degrade insurgent operations in the area.

When Kara found her target, she would brief the team that would be tasked with taking the enemy out. On one occasion, after she excitedly found the HVT she was looking for, a SEAL team was assigned to raid a local village and complete the mission.

As she briefed the team, one of the SEALs stood in the back, arms crossed, delivering a dark, disapproving frown worthy of a Clint Eastwood close-up. He told Kara she was dead wrong. She was going to get them killed.

Kara knew she was right—except she wasn't.

The mission was scrubbed. Indeed, the target they were looking for wasn't there. The annoying, know-it-all SEAL knew what he was talking about.

Unfortunately for Kara, that wasn't the last she saw of him. Afghanistan is a big country. The outposts where Americans worked were not. Whenever the SEAL ran into her, he couldn't help reminding Kara of the day she almost got him killed. This apparently was the basis for a blossoming friendship. You could see why. Kara is whip-smart, self-confident, and fearless (not to mention a talented athlete who had played Division I soccer in college), more than a match for a SEAL. Friends turned into friendly dating. Years later, that turned into marrying.

To this day when you ask Kara how she met her husband, she starts, "Well, I almost got him killed." This will no doubt make an interesting story for their grandchildren.

Even though Kara wasn't always right, she was an extraordinarily gifted analyst with an intuitive skill for mapping insurgent networks. If she was playing professional baseball, her batting average would have gotten her to the all-star game.

Still, although tracking terrorists is rewarding work, it is not a career, not to mention that three Afghanistan deployments are enough for anyone.

Kara was recruited by Facebook to help fight extremism online. This opportunity kept her in the fight but with much better working conditions than the testosterone-laden, sweaty stench of command posts in Afghanistan.

As discussed in chapter 2, over the last decade, violent Islamist extremist groups made a concerted effort to establish a powerful presence in social media for fundraising, recruiting, and inspiring terrorist attacks like the Boston Marathon bombing undertaken by the Tsarnaev brothers (discussed previously). All the major tech platforms found themselves having to dedicate more and more resources to combating not just criminal or malicious activity online (like YouTube battling bots) but also actions that could get people killed.

There were lots of reasons why Facebook would want to hire someone like Kara. She had real-world operational experience. She is also a big brain. The Marine Corps brat had a BA in foreign affairs and history from the University of Virginia and an MA in war studies from King's College London. She had all the right stuff, all the skills, knowledge, and attributes of a real network warrior.

Kara helped create Facebook's Global Security Counterterrorism Analysis Program. She was also the team lead for the Facebook Headquarters Regional Intelligence Team in Menlo Park, California. Her expertise was in mapping actions online, identifying activities that might lead to people getting exploited, hurt, or killed. This was the kind of oversight that if it had been in place might have stopped someone like the Tsarnaev brothers.

After a few years as a California girl, she was recruited again. This time, she departed the world of "big tech" for the think tank universe. In part, Kara left the shadow of Silicon Valley because she learned there were parts of the tech world that she was not happy with. It was one thing to analyze networks to stop malicious actors online, but it was another thing to use the power of the platform to play politics (more on that in later chapters). Kara Frederick decided she needed space that would give her a more independent voice in what big tech was doing right and wrong. She morphed into a powerful civil society umpire calling balls and strikes over how tech companies used and abused their power.

Frederick quickly became not only one of the leading US analysts on tech policy but also one of the more well-known ones. She is a regular guest on Fox News and Fox Business. She was interviewed on CBS's *60 Minutes*. She pops up often on all kinds of other national and international media and podcasts. She is prolific as well, publishing in the *Wall Street Journal*, *Foreign Affairs*, *Foreign Policy*, and elsewhere.

In 2025, her expertise and talent earned her a spot on the US president's National Security Council Staff. Now she really worries the bad guys.

What made her a hot commodity in Washington is her knowledge of the many worlds on the internet from the dark web to inside terrorist cells to the practices of big tech and the activities of global actors like China and Russia. If you ask her how do you become the master of these different spaces, she will tell you it all starts with knowing the players and what they are doing—and that knowledge comes from mapping the network.

Mapping

The examples of Kara Frederick and the Joint Improvised Explosive Device Defeat Organization highlight the importance of mapping networks if your intent is to take them down. Still, my favorite example is *Donnie Brasco* (1997). The film is based on the true story of undercover Federal Bureau of Investigation (FBI) agent Joe Pistone who infiltrated a Mafia crew. Joe was charged with helping the bureau map out the scope of the vast criminal enterprise the crew supported. In the film, Donnie gets recruited by low-level gangster Lefty Ruggiero. Lefty reveals to Donnie how things work. "If I say you're a friend of mine, that means you're connected," he explains. "If I say you're a friend of ours, that means you're a made guy. If I introduce you, I'm responsible for you. Anything wrong with you, I go down." This advice was just the start. Tutoring Donnie, Lefty offered up the mob's crown jewels, describing how the network worked. Donnie understood the importance of this knowledge; so did the FBI. Now they knew how to take Lefty and his friends down. Starting with low-level arrests, informants, and plea deals, they worked their way up the chain of command to identify, gather evidence against, indict, and convict the capos and bosses managing the Mafia families.

Indeed, US law enforcement has developed the mapping of criminal networks into a fine art. One of the penultimate examples from the cyber world is a case mentioned earlier—the Silk Road. In 2011, an informant told investigators at the Department of Homeland Security, "You guys are in law enforcement. You might want to look at this."[1] That started something. The Silk Road, it turned out, operated on TOR, short for the Onion Router, an open-source software used for anonymous communication, popular with malicious actors who want to operate on the dark web. The Homeland Security investigators were stunned at what they found on TOR, an emporium that peddled everything from hard drugs to murder for hire.

Part of the effort to go after the Silk Road started when "law enforcement in Maryland began mapping the operation. They focused on identifying and nabbing two groups connected to the Silk Road, "the top 1 percent of sellers and the moderators and system administrators, whose computers and credentials, once seized, could open the door to the site's private communications and account details." Just like the FBI drew a wiring diagram of the Mafia in the real world as a precursor to putting people behind bars, law enforcement followed the electrons online to map out the virtual criminal enterprise behind the Silk Road.

Joe Pistone's undercover operation eventually led to the takedown of cigar-chomping, overweight bigwigs. When they followed the route of the Silk Road, it led to a real-world Ernst Stavro Blofeld (the fictional criminal mastermind behind James Bond's nemesis SPECTRE) directing the whole criminal enterprise. He was the Dread Pirate Roberts (that's right, the character from the 1987 film *The Princess Bride*). The Dread Pirate Roberts was— wait for it—a twenty-nine-year-old computer nerd with delusions of grandeur named Ross William Ulbricht. Last time I checked, he was still behind bars.

If the plan is to undertake social networking warfare, what is the alternative to mapping networks? Reflect on the previous discussion of artillery tactics in Vietnam, where the Americans hurled round after round into the inky black jungle night hoping to hit the bad guys. The United States expended enormous effort inefficiently because they didn't know what they were looking for. Likewise, before Donnie Brasco, even when the cops arrested a mob figure, all they had on their hands was one mobster. The Mafia's famed code of silence, the omertà, meant death for any member who ratted out the other mobsters to the cops. This was the Mafia's firewall. It stopped the FBI cold.

The alternative to not knowing who you are going after is to map the network. For instance, Lefty Ruggiero (real-life friend and mentor of Donnie Brasco) unknowingly helped open the whole criminal network to the FBI. Thanks, Lefty; hat tip to the Dread Pirate Roberts as well. There could not be better examples of the difference mapping can make in network warfare.

Mapmaker

In taking down the mob, the FBI tracked orders. Who was connected to whom? Who gave orders and who followed them? On the Silk Road, law enforcement tracked orders, too—orders for goods and services. Social network mapping is identifying and interpreting interactions on the network between individuals and groups to determine the level of influence. What users exert the most influence over the other members of the network? Understanding the flow of influence is the compass of social network mapping.

Mapping social networks can be challenging. Unlike knowing which thug reported to which mobster or who bought drugs from whom on the Silk Road, social networks, particularly as they scale, can be complex and nonlinear where it can be difficult to sketch the paths of influence and action. Mapping networks can become even more complicated when networks overlap and influence one another, interactions that can occur often when the real and virtual worlds connect and users operate in multiple domains.

Mastering Complex Systems

Anybody can probably figure out how to map a social network if it is two people chatting about where to go for lunch on WhatsApp. On the other hand, for bigger, messier networks, assessments can get a lot more problematic. To address that challenge, social network mapping draws a good deal from the practices of complex systems analysis.[2]

A system is "any set of regularly interacting factors and activities that has definable boundaries and that produces measurable outputs." The complexity of a system is determined by the number and diversity of interacting components. When systems become overly complex, their behavior cannot be easily predicted by traditional methods of analysis, like breaking an assembly line into its component parts and analyzing elements in detail as Frederick Taylor did when he perfected scientific management.

In a complex system, elements are so interconnected and their relationship so multifaceted that their properties cannot be properly understood without assessing their interrelationship with each other as well as their relationship with the wider system and its environment. Financial systems offer a case in point. It is often difficult, for instance, to appreciate the value of an individual stock just by knowing its current price. On the other hand, when the price of the stock is understood in relation to the performance of the larger system (e.g., the market—whether the average price of stocks on the exchange are on a positive trend [a bull market] or on a negative trend [a bear market]), the value of the stock takes on a more significant meaning. In short, in a complex systems analysis, assessing individual properties often requires mapping them according to their place in the overall system. Assessing influence in a social network can reflect the same kind of challenge.

Mapping complex systems can be maddeningly difficult. Complex systems can exhibit patterns, outcomes, and properties that are not present in any of their individual elements. This is so super frustrating. Therefore, the performance of these systems is often described as nonlinear, meaning that how the system functions cannot necessarily be derived just from understanding the sum of the behavior of the many parts that compose the system. Financial markets again offer an example. Analysts can study all kinds of information on the components of a market, from the monetary exchange rates to interest charges, yet cannot accurately predict whether future markets will be bullish or bearish. The same unpredictable behaviors can happen in complex social networks.

Complex National Security

The consequences of mapping networks effectively is a big deal when matters of national security are on the line. Since complex systems are so much more difficult to analyze, understanding how they work, predicting their behavior, or determining the optimum means for changing their performance presents a unique challenge. When the performance of systems affects the security of the nation, the task can be particularly crucial and daunting.

Many national security problems involve attempting to understand, predict, or affect the behavior of complex systems from national emergency response to transnational terrorist organizations (like the ones Kara Frederick tracked in Afghanistan). Failing to understand how discrete decisions impact the whole system can produce unintended and counterproductive consequences. Here is an example. In the aftermath of Hurricane Katrina, emergency officials barred all but authorized emergency responders from entering New Orleans. As a result, fuel handlers who had not been credentialed by state officials could not make necessary deliveries to generator-powered emergency centers. Without gas or fresh batteries, the centers lost power and became inoperable.[3] Since officials failed to understand how the entire system worked, they fixed one problem, preventing unnecessary convergence at the disaster scene, and created another—disabling key command and control nodes. That is the kind of outcome network warriors should seek to avoid. When messing with a national security network with the fortunes of nations and life and death on the line, dominant networkers ought to be able to act with confidence, knowing how their actions will affect the system they are messing with. That is why mapping is so important.

Complex networks can be hard targets. Some terrorist groups, for instance, have demonstrated tremendous capacity for adaptation. According to a report by the Homeland Security Advisory Council, a terrorist group can be "proactive, innovative, well-networked, flexible, patient, young, [and] technologically savvy, and learns and adapts continuously based upon both successful and failed operations around the globe." Some terrorist groups may not only be best understood as complex systems; battling them may also require understanding how these groups exploit other complex systems such as the internet. Bottom line: The ability to effectively analyze complex systems has utility both for safeguarding against threats as well as mitigating or defeating potential or existing dangers.

Mapping Complex Systems

Describing complex systems—how they work, what they produce—and then determining how systems performance can be changed (for better or worse) is the task of complex systems analysis. Until recently, researchers have approached complex systems by means of traditional means of quantitative analysis—breaking them down into their smaller constituent parts and analyzing these in detail, just like Frederick Taylor did with pig iron. Cell biologists, for instance, study organisms in terms of how their component cell systems interact. Similarly, engineers try to understand how a vehicle operates by taking apart all its individual components, analyzing their properties, and reassembling the vehicle in a slightly altered way to observe the effects on its performance. That mechanistic approach doesn't work as well with complex systems like social networks because it is often just not clear how piece parts go together.

It is not just a problem of not knowing enough about what comprises the system. For example, you might have a list of every person in a social network and still not understand how to map influence in the system. Complex systems are more than a counting problem. There used to be a working assumption that as data collection and processing techniques continued to improve, researchers would be able to develop superior models for predicting system properties, even very complex ones. But contemporary research reveals that it is not just about a lack of information to model or analyze the system. Standard system models remain deficient because they still offer oversimplified descriptions of system behavior. Simply adding more data to the study of a complex system may not be enough. In contrast, complex systems analysis tries to bridge the gap between knowledge and understanding by exposing and studying interrelationships. That requires a more sophisticated mapping process.

Fortunately, there is a lot of research on complex systems analysis. People have been grappling with the analysis of complex systems long before Vint Cerf helped invent the internet. Complex systems have been studied in the natural world for decades. Ecologists have applied this type of analysis to the life sciences using mathematical models to help understand what drives large fluctuations in wildlife populations. They have also used computer models to establish the properties of small-scale systems to identify emergent properties of molecules within cells. Often, the complexity of the system of interest requires employing a multidisciplinary approach that combines insights from several scientific fields.[4]

The practice of mapping complex networks is well established. Maps of networks are constructed with "nodes" and "links." A node is an entity, like a user on a mobile digital device. The link is the connection between nodes. These are often represented by dots and lines. The shape, size, and number of nodes and links visually depict how connections are structured within the network.

To gain an appreciation of the dynamics of a complex system and the amount of activity (in the case of social networks, influence), analysts use a model composed of "stock" and "flow" diagrams. A "stock" represents an entity (e.g., money) accumulated over time. Stocks change through "flows." A flow is a change in stock over time (e.g., deposits and withdrawals from a bank account). The relationships between stocks and flows in a complex system are depicted graphically through how they move over the network of nodes and links. The flow of activity and influence over social networks can be portrayed in the same manner.

Mapping Tools and Techniques

If this sounds a bit daunting, that is because it can be. People get a PhD and fire up supercomputers to do this stuff. That said, undertaking this task is not out of reach for anyone. In 2021, the website Rank Red, for example, lists twenty-three open-source, free online tools for social network mapping that provide means to "make it easier to carry out qualitative or quantitative analysis of social networking platforms. They describe the network's attributes and workflows via visual or numerical representation."[5] These free programs are pretty powerful, and they are just the bottom feeders of the analytical tools that are out there. There are tools for virtually every level of competitor in the online social networking space.

Also, don't get the wrong impression. It is not always necessary to be a data scientist to map social networks. Even without a computer, mapping a network isn't beyond the reach of any network warrior. Mapping is a process. The more tools you have, the more you can do. Pioneers can build a log cabin with an axe. Developers can build skyscrapers with an army of skilled workers and machines. The axe might be enough, or you might need a work crew. It depends on the nature of the competitor you are facing and how important it is to understand the inner workings of a network to compete with them.

Regardless, if the task is taking down the Silk Road, disrupting the command and control of Russian nuclear systems, or manipulating a social network, the process of mapping has some common elements. Four are described below.

Bounding the System. A system is defined by the external environment and the internal components of the system. Inputs come from the external environment. Outputs are what the system puts into the external environment. Defining how to bound the system can make a big difference in shaping the scope and complexity of the mapping tasks. For instance, smelting iron includes everything from mining and transporting to all the activities that happen at the smelting plant. Frederick Taylor bounded the challenge of analyzing operational activities by just looking at the process of taking the pig iron off the railcars and getting them to the smelter. Bounding the scope of the system that he was looking at made the challenge of analysis more manageable. In comparison, look at the difficulty the Chinese government made for itself when it decided to monitor all online activity in China through the great firewall treating the entire online population as a single system (discussed in chapter 2). No wonder their efforts are so inefficient.

How do you decide how to bound the network you want to map? The answer goes back to your mission. What are you trying to accomplish? The scope of the system must be large enough to impact the outcomes you are trying to achieve. During World War II, for example, the army air forces wanted to cripple the German industrial base producing fighter aircraft. Therefore, they had to map the whole German industrial base that produced fighter aircraft as a starting point before they decided what key targets to go after in Operation Pointblank.

Data Collection. There is an adage: Quantity has a quality all its own. It is impossible to map a system without sufficient data. The Silk Road, for example, was just a mysterious criminal conspiracy until law enforcement officers collected enough data to map the network. That was initially a big challenge since the network was operating on the dark web using encrypted information. Without finding a way to get access to the data, the police investigation would have gone nowhere. If you can't get the data, you can't map the network. So data availability is a huge factor in determining what can or cannot be done.

There are several challenges to data collection. One is that your adversary might try to prevent you from getting the data. In every mobster movie anyone has ever seen, there is always a scene where a fat guy crams into a telephone pay booth (back when such things existed) to "talk to a guy about the thing." The mob knew that the FBI made a practice of tapping phones, so mobsters always tried to find ways to ensure their conversations were not overheard, or they would speak in generalities that if the calls were tapped, they couldn't be used in court as evidence of criminal activity. What were they doing? They were trying to deny the FBI data that could be used to map the network.

Another issue is, is there sufficient data to be collected? I have been an adjunct professor for many years. My favorite course is teaching methods of research. I teach that by challenging students to come up with research topics. One of my favorites that they often come up with is, "How does the North Korean government make decisions?" I love that question because I get to ask the students, "What are you going to use for data?" The North Korean regime is the most closed and opaque governing body on the planet. I mean, it is not like they are posting their cabinet meetings on Rumble (a video-sharing site). You might as well ask what's up in the Star Chamber (a secret or closed meeting held by a judicial or executive body). If there is not sufficient data to map a network, mapping is kind of a useless exercise.

If there are actual data to be gotten, there are lots of ways to get them, depending on your needs and resources.

Some data you can get voluntarily. For example, you can just ask people. Surveys are a common tool. Applications like SurveyMonkey is one example of an instrument that might be helpful for collecting data.[6]

Some data are "open source," which means they are publicly available. My think tank runs a project that collects vast amounts of data on China by collating open-source databases and studies on Chinese activities (see chapter 7).[7]

Some data you can buy. For example, Amazon will sell you access to their AWS Data Exchange, which will give you admission to one of the world's largest collections of publicly available third-party databases.[8]

Some date you can obtain illegally. Buying stolen data is totally a thing. Even law enforcement agencies buy stolen data, though they usually use it to help track down the hackers who stole it.[9]

Some data can be obtained through covert action or intelligence gathering. In World War II, in Bletchley Park, British cryptologists decoded some of Germany's most secret message traffic. This information was useful for mapping the scope of German military forces and operational practices.

Process Tracing. Once you have data, it is almost impossible to map networks without some means of process tracing. As one scholarly article explained, this is "the systematic examination of diagnostic evidence selected and analyzed in light of research questions and hypotheses posed by the investigator."[10] Put simply, this means being able to demonstrably identify a connection between two entities on an issue that you are interested in. Here is an example. At the Battle of Midway (June 4–7, 1942), which has been mentioned before, the chief of naval operations (B) dispatched a fleet to intercept the Japanese force at Midway because the US intelligence (A) obtained key information that the Japanese were planning an attack on Midway. How do we know A influenced B for sure? Because we have the documentary evidence to prove it. In short, process tracing is having the ability to draw causal inferences from available data. If you don't have a way to do that, you can't map a network.

Research Design. Rather than being like the US artillery firing blindly into the night, your odds of mapping adversarial activity go way up if you have a plan to collect, analyze, and interpret the information needed to map a network. In my teaching days, I called this the "research design," the elements that needed to be in place before one could confidently go forward with a research project, like mapping a network. This requires putting together some of the key tools and techniques already discussed. Together, they comprise a useful framework for organizing many kinds of research and investigative activities.

There are three crucial elements to a complete research design.

Question. Do you have a relevant, important, and answerable question? In this case, can you reasonably expect to identify the influences that are important and can be measured? Let's go back to the example of mapping the German industrial base and the genesis of Operation Pointblank. Mapping the German industrial base made no sense if after the exercise the planners could not answer the question, what was the center of gravity for German fighter aircraft production? This knowledge would pinpoint the key target for the bombers to go after. Without that answer, the map was useless. You may not need to know everything about a network to map it, but you must know what you need to know.

Data. There is no use trying to map something without data, particularly if you can't get data that can answer the question. One of the reasons Operation Pointblank failed was because the Allies had no clue as to how many ball bearings Germany had stockpiled. The answer turned out to be enough to make up for the production lost from the Allied bombing of the German ball bearing plants. Since they could not get the crucial data they needed, in the end their map was not much help. If you can't get the data you must have to answer the question, Houston, you have a problem.

Method. If you cannot analyze the data you have in a manner that allows you to answer the question, your mapping exercise will prove fruitless. The British intercepted lots of German encoded transmissions. For the first part of the war, the intercepts just piled up. The problem was, until the Allies discovered a method to decode the secret messages and utilize the information, they couldn't map anything.

So that is the real magic of mapping networks. Before you start, make sure you have the question, data, and method issues addressed. Otherwise, you won't be able to map squat.

Technology

There are tons of technology tools, with new ones appearing all the time, that might be useful for mapping social networks. Some have been previously mentioned. One described in a preceding chapter was link analysis software. There are all kinds of techniques and algorithms and software that can be used for extracting information from network data. You can learn how to do this.[11] You can also pay others to provide you the service. Microsoft Analysis Services, for example, offers platforms and services for a price.[12]

Other technologies offer innovative ways to map networks including using mobile devices and the Internet of Things (more in chapter 14) to track human interactions online and offline. One research team built an experimental Human Atlas to do just that. "Most social network analyses focus on online social networks, such as Facebook, Twitter, and LinkedIn," the research team noted. "While these networks encode important aspects of our social lives, they fail to capture many important real-world connections."[13]

So what this team did was build a web-based tracking tool that mapped the social network of the MIT Media Lab in both the physical and virtual space. In another experiment, researchers used "smart badges" to map social networking behaviors at a professional conference.[14] What this research suggests is that in addition to the technologies explicitly designed to provide data collection for mapping purposes, virtually any technology that collects and stores digital information for any purpose could potentially be harnessed as a source for social network mapping. More on the implications of this in a later chapter on the impact of emerging technologies (chapter 14).

Offense and Defense

Fighting a modern war without a map is unthinkable. That holds true for networking warfare as well. Having a map, however, doesn't help much if the map is not used to maximum affect. Napoleon loved maps. He had a corps of geographic engineers to make them for him. He took his own map wagon on campaign to plot his operations. As noted, maps aside, Napoleon got his butt kicked on the Russia campaign (1812), as Carl von Clausewitz took pains to remind us. Reading a map is not enough.

Great warriors, including network warriors, must learn to make the most out of the maps they have, whether on offense or defense.

This reminds of another military practice that has suitable application for network warriors. Remember the example cited of the United States wasting artillery in Vietnam? After Vietnam, the army became downright fanatical about doing better. The generals wanted to make sure that they did not waste assets like artillery fire. Indeed, they wanted to make sure the army used their firepower to maximum effect, putting rounds on the most important targets, at the most important place on the battlefield, and at the most important time. So the army developed a means to do that based on analyzing maps to predict enemy actions and how to mess them up. This practice was called the Intelligence Preparation of the Battlefield (IPB). It is still in use today. In fact, IPB was part of the tools Kara Frederick used to track down terrorists in Afghanistan.

The military really didn't invent anything new. What folks did was combine timeless sound military practices to form a logical and intuitive way to think about battle before the battle and then give it a fancy new name that made them sound smart.[15] Still, the IPB offers a structured, disciplined tool for analyzing maps as well as the other stuff that shapes how competitors compete in physical as well as digital spaces. So it is worth looking at how it works.

There are four steps to the IPB process: (1) Define the battlefield environment; (2) Describe battlefield effects; (3) Evaluate the threat; (4) Determine threat courses of actions. Let us run through and examine how these four steps have application to social networking warfare.

Define the Environment. This just means knowing as much as possible about where you are operating. In combat, this includes the geography, weather, people—pretty much any force or factor that might affect how military forces would operate in the battlespace. For example, when the Allies planned their operations in Normandy, they spent a lot of time studying the tides and the weather because they knew that these would significantly impact their ability to conduct amphibious operations—getting troops from ships at sea to the beaches. This assessment process is also used to identify key information that is not known, so you can then think about how to get that data, too, or how to account for the absence of that knowledge. IPB is an equally useful method of inquiry if you plan on fighting a digital networking war. You must know the digital terrain from the internet service providers to the capabilities and limitations of platforms. Such a step, for instance, was an absolute precursor to the law enforcement operation that went after the Silk Road.

Describe Effects. Evaluate how the environment will affect the operations of both sides. For instance, when the Allies were planning the invasion of Normandy, they asked citizens to send them all their postcards from their holidays on the French coast. Why? Were they thinking about postwar vacations? No. They were looking at every photograph they could get their hands on to figure out where there were suitable spots to land an invading army. This practice makes just as much sense in the virtual networking space. For instance, before they could take down the Silk Road, law enforcement officers had to understand how the use of encryption technologies impacted the criminal's ability to operate and how the cops could access encrypted data so that they could go after the bad guys.

Evaluate Threats. What can the adversary do, where, and when? No adversary can be everywhere all the time, doing everything. They must decide where to attack or defend. So combining the knowledge from the first two steps and understanding the enemy's capabilities reveals where, when, and how they can operate. The invasion of Normandy was preceded by a massive intelligence effort to identify what German forces were available to fight in Normandy. The decryptions from the intelligence units at Bletchley Park, called "Ultra," were a huge help in this regard. In another example, once the investigators got inside the Silk Road, they began to gather similar information, understanding the scope and nature of the criminal enterprise they were up against.

Determine Course of Action. Armed with the insights gained from steps one through three, you can make assessments of what are the most "likely" and most "dangerous" actions your adversary might take. The Allies followed this process to help determine that they had the best chances for a successful amphibious landing in France by hitting the beaches in Normandy in late May or early June 1944. With that knowledge, they planned all kinds of countermeasures to undermine the German defenses. That didn't work out too bad. In a similar fashion, a task force of investigators used their extensive analysis of their Silk Road mapping to plot a prosecution campaign that eventually took down the whole criminal network, an outcome that proved dreadful for the Dread Pirate Roberts.

In short, what the IPB framework provides is a way to think proactively about how to use network mapping for offensive or defensive operations. This is about as handy a weapon of networking warfare as one could ask for. The IPB is to winning warriors like dance moves are to the Temptations.

Attack!

This chapter summarized the precursor to all activities for taking serious action against an adversarial network. That starts with gathering the skills, resources, and data to map the network. With a map to guide future efforts, you can then start to conceptualize a campaign against your enemy using the map to do an assessment of actions and counteractions that might unfold (through an IPB or similar process) in the competition. This foundation prepares you for the next step—deciding what actions to take and how to implement them.

Now armed with the skills of how to map social networks and the means for turning that knowledge into proactive action, we have reached the appropriate point to turn to how to make network war. That discussion, the instruments of mayhem and marauding, starts next. In the chapters ahead, *Digital Dominance* looks at how to disrupt, defeat, and destroy social networks.

8

KEEP THE TOWELS

Seb Gorka is the definition of a disruptive influence.

Love him or hate him, there is no denying that Sebastian Gorka knows a lot about disrupting online and offline. Seb has been on both sides of this fateful coin toss. His career as a presidential adviser was cut short by a concerted campaign to rattle the office of the newly elected president. After he crashed and burned on the White House lawn, Gorka became an influential voice on talk radio, social media, broadcast news, and political networks harassing and savaging the forces that came after him. Harnessing a lot of what he learned as a military analyst and counterterrorism expert, Gorka has launched an impressive social networking counteroffensive of his own.

How did Seb Gorka wind up with a target on his back? Well, if you know how Washington network wars work, this would come as no surprise. If you do not, there is a backstory.

Having known Seb for over a quarter of a century, in all those years, I would never have predicted he would become either a lightning rod of controversy or an activist icon, but such is the power of networking.

Earlier in *Digital Dominance*, I mentioned I had a penchant for discussing military matters because I came from a military background (albeit also because military competition has a lot of application to social networking warfare). While I was in the army, the last six years were pretty much un-army time. I was stationed in Washington, DC, about as far from places where boots get muddy as one can get. Not much saluting. No shooting. All the warfare in Washington is mental. In six years, I saw a world war's worth of that kind of combat.

For part of that time, I was the speechwriter for the army chief of staff, the senior officer in the service. In fact, "chief of staff" was the job George

C. Marshall had during World War II. I was constantly reminded of that. His desk was enshrined in a glass case down the hall. While I had no rank or position of note, this was a perfect perch for observing the battle of Washington powerhouses—generals, politicians, and bureaucrats lobbing their opinions and influence back and forth across the Potomac like a barrage preceding the troops into battle. At the same time, I myself crossed the Potomac a couple of days a week to finish a PhD at Georgetown University. After the chief retired, I decamped permanently to the other side of the river to work at the National Defense University at Fort McNair. There, I edited a journal that Colin Powell started when he was chairman of the Joint Chiefs of Staff called *Joint Force Quarterly*. When I finally retired, I just stayed in our nation's capital and went to work in think tanks—first, a small defense think tank for a year and then my present perch ever since.

I mention this history only because in all those years, I was constantly bumping into this Sebastian Gorka guy. At one point, we were both adjunct teachers at the same graduate school. I wound up substitute teaching his course on Islamist terrorism for a semester (to be honest, they turned out to be big shoes to fill).

So who the heck is Seb Gorka and why are they saying all those terrible things about him?

For starters, to understand Sebestyén Lukács Gorka, you must understand his father, Paul (Pál). Paul was born in Hungary. In 1940, Hungary joined the Axis powers. That didn't work out so well. Hungary was devastated under the Nazis and then invaded by the Russians, who after the war installed a brutal Communist dictatorship.

Paul didn't like the Communists any better than the Nazis. He organized a secret student resistance movement. That didn't work out so well either. His network was betrayed by the infamous British double agent Kim Philby. The Hungarian Secret Police arrested Paul. That really did not work out well. He was tortured and given a life sentence.

Paul spent years in solitary confinement. Then, for vacation, he was sent to work in a brutal prison coal mine, then moved to the main political prison in Budapest. Believe it or not, that was Paul's lucky break. In October 1956, the Hungarian Revolution broke out. The first thing the revolutionaries did is what revolutionaries always do. They liberated all the political prisoners. Paul's freedom looked short-lived. Things went not so well again. The Russians returned and squashed the revolution like a grape in a power press.

This time, Paul escaped to Great Britain. On the way, he helped the seventeen-year-old daughter of a fellow political prisoner. They married. Paul and Susan (Zsuzsa) had a son. They named him Sebastian.

This backstory is important because Seb learned two things from his parents. One was a love of freedom and a passion to oppose its enemy: authoritarianism in all its forms. The other was he learned how to fight back. His parents taught him to be a thinker and a fighter. That meant being smart and tough.

Gorka received a bachelor of arts degree from the University of London. While at university, he volunteered for the British Territorial Army, the rough equivalent of being in the National Guard in the United States. This was heaven for Gorka. He got to play with his two favorite things—books and guns.

When the Cold War ended (1989), the promise of the Hungarian Revolution was realized. Hungary now had democratically elected governments. Gorka moved to Budapest to support the new democratic regime and continue to pursue academic research. He worked for the Hungarian Ministry of Defense and did a master's degree at the Budapest University of Economic Sciences and Public Administration (later renamed Corvinus University). He dabbled in center-right politics and later pursued a PhD at Corvinus University.

Gorka also met an American girl from Pennsylvania. They got married. More on her, the "really dangerous" Gorka, later. Together, they founded a think tank, the Institute for Transitional Democracy and International Security. Among the issues they focused on is transnational terrorism.

On 9/11, Gorka became an overnight minor Hungarian television celebrity. He was one of the few people in Hungary who could spell al-Qaeda. When the news of the attacks on New York and Washington broke, he called the station to complain about the coverage. Did these people know nothing? A producer answered the phone. They said, "How soon can you get down here?" He was a fixture on Hungarian television for weeks.

His expertise on terrorism led to a position teaching at the George C. Marshall European Center for Security Studies in Garmisch-Partenkirchen, Germany. Yes, named after that George Marshall. The Marshall center, cofunded by the US government, introduced Seb to American military and academic circles. The Gorkas emigrated to the United States in 2008 (later, he became a naturalized American citizen). The Gorkas had two children. They wanted them to grow up in the world's epicenter of freedom—the land of the free and home of the brave.

Seb got a job at the National Defense University in Washington, DC. Our paths have been crisscrossing ever since. I don't remember how he became an adviser on the presidential campaign. I did not run in those circles. I worked on the transition team for the presidency, which was a strictly nonpartisan, nonpolitical effort.

When his candidate won, Gorka ended up in the White House as a presidential adviser. Then the knives came out. The attacks on Seb were merciless and relentless. He even earned a profile in the *Rolling Stone* that accused him of being a racist and a Nazi, among other things.[1] Gorka can defend his credentials and his record himself. But I can, and want to, say two things here on the record. One is that he is not a racist. That was a completely manufactured canard. Opposing Islamist terrorism doesn't mean you hate Muslims. It means you hate terrorists (most victims of transnational terrorism, by the way, are Muslims). Two, he is not a Nazi. The haters think anyone who embraces center-right politics is fascist. But Seb was and remains his father's son. Paul hated Nazis and Communists with equal fervor. How do I know these things? Because I know Seb.

I cite Seb here not because of his personal trials but because Seb as a world-class security analyst would recognize and appreciate exactly what was being done to him. One way to fight a system, any system, any network is to disrupt it. Seb Gorka became a disruption for the White House. In the end, he had to leave.

Seb's wife, by the way, went to work for the Department of Homeland Security. She too became a target.[2] She was even described in the press as the "really dangerous" Gorka.

That, however, is not the end of this story. Before joining the president's team, Gorka had been working for a prominent social media news outlet heading up reporting on national security. After he left the White House, he put those skills and his media experience to use. He became a prominent voice on talk radio, social media, and broadcast news.

It doesn't take much watching Gorka in action to see that he has put all his expertise in warring to work—and he is very good. He is a master in the art of disrupting powerful people with inconvenient truths. Love him or hate him, he is a fighter.

In 2025, Gorka was asked to return to the White House. Today, he oversees the administration's global counterterrorism operations—proof enough you can't keep a good guy down.

Gorka's experiences in Washington in fighting are a great case study in networking warfare. There is lesson here. Sometimes, all that is needed to win is to disrupt your adversary and put them off their game.

Disrupting Networks

The objective of disruption is to create enough disturbance and interference in your competitor's network to keep them from accomplishing their goals or creating a space, while they are in chaos, that allows you to achieve your goals.

How does disruption work? Let's go back to the movies. Personally, I thought *Ocean's Eleven* (1960) was a terrible motion picture, an excuse for Frank Sinatra and his Rat Pack buddies to drink on the clock. The film was certainly not worthy of a remake with an all-star cast led by George Clooney.

To be honest, I don't think the 2001 version was all that much better. It did, however, have a fantastic scene that explains exactly what disruption tactics are all about.

When George Clooney meets his rival, a powerful casino owner (Andy Garcia) whom he and his gang intend to rob blind, the casino mogul warns George, "I know everything going on in my hotels." He narrows his eyes with a "I am watching your very move" sneer. George's character, Danny Ocean, quips, "Then I better put the towels back." A smug-grin Garcia replies, "No, you keep the towels." This little clip reveals exactly how Danny is going to win. In movie lingo, they call this foreshadowing. Here is how we know Danny Ocean is going to come out on top.

The security at the casino was premised on continuous, persistent, total surveillance that allowed for knowing anything and everything going on everywhere all the time. Ocean's eleven (the number of thieves in the gang) figured out a way to beat the security. Spoiler alert—The scam works. You know why? All you need to do is disrupt people who think they know everything and make them continue to think they are right, especially when they are wrong. Don't think because you can count all the towels that you are safe. If an opponent can find a way to disrupt the network, the purpose and the action of the network can be put at risk. That's what Danny did.

Here is a real-life example of how disruption works. Let's return to the Normandy invasion during World War II. One intractable problem the Allies faced was that even if they got to the beaches, the Germans might counterattack and drive them into the sea. If you want to know what a debacle like that looks like, read the history of Operation Jubilee, the Dieppe Raid, an amphibious attack on the German-occupied port of Dieppe (1942). It did not end well. The Allies could not risk a failure like that again.

Adolf Hitler did them a favor (not on purpose; he thought he was being clever). Rather than positioning his counterattack forces right near the beaches, he concentrated them miles from Normandy. This meant that the Germans had to wait until they were sure where the Allies' main attack was happening before they launched a counterstrike. The Allies, who were aware of the disposition of German forces, knew that they needed to find a way to delay the German counterattack until the Americans were firmly rooted on the Normandy coast and could fight back.

The Allies came up with a couple of ideas. One was called the Fortitude deceptions. They had two plans, Fortitude North and Fortitude South, that created fake armies to convince the Germans that the main Allied attacks would occur in other places, hoping Hitler would think the Normandy invasion was a ruse intended to draw the Germans away from the main effort. Furthermore, the Allies had plans to slow the German reinforcements when they were dispatched to Normandy. One effort called on French resistance fighters to conduct acts of sabotage. Another initiative was the Transportation Plan: Allied bombers would pummel bridges, railroad yards, and railways across France before D-Day so that they would be unusable for the German counterattack forces. Finally, the Allies dropped airborne forces (the same guys who would later train Bob Goldfarb) around the Normandy countryside to fight off any German counterattack forces that showed up while the Allies were landing on the beaches.

The cumulative effect of all these actions, as well as the fact that Hitler overslept on D-Day (and that the commander of the Normandy defenses was in Berlin buying his wife a pair of shoes for her birthday) created enough disruption to allow the Allied landings to succeed.

In contrast to Vladimir Putin's Russia which often blindly lashes out with disinformation and other disruptive campaigns to annoy, frustrate, and undermine democracies, targeted, dominating action like the Allied invasion of Normandy worked because the Allies conducted a process like the Intelligence Preparation of the Battlefield (IPB) framework described in the last chapter. An IPB-like method helps determine where, when, and how an enemy might be vulnerable at the decisive moment you need an advantage. You must have that knowledge if you plan on proactively annoying your adversary in a serious way. That's why, when I began to write this chapter, Seb Gorka was the first person I thought of. As a trained strategist, I know of no one who knows better what can be accomplished when these practices are employed to maximum effect.

Menu of Mayhem

How much disruption do you need to impact a network's performance sufficiently to achieve your objectives? The answer is simple: as much as is needed. It is all relative to the situation. What is your enemy trying to accomplish? What are you trying to accomplish? What are the forces and factors that impact system performance?

The strategist John Boyd, who was discussed in a previous chapter and profiled in *Wiki at War*, is most famous for his observe, orient, decide, act loop model of competitive decision-making. This model has great utility here. If disruption is the aim, then the goal of the interference effort is to slow your adversary's decision loop enough so that you can act faster and more decisively. The key to disruption—and this is important—is that it is not just about frustrating the adversary. It is also about planning an action or activity that takes advantage of the disruption to gain a decisive advantage. In the case of the Normandy invasion, for instance, the Allies just needed to slow down the German high command enough so that the Germans did not decide and act to commit the reserves before the Americans were ashore and could defend themselves from a concerted counterattack.

Exploit the Network. Before disrupting a network in some cases, information can be gleaned from the network to help the competition, or the adversarial network can be manipulated to shift action in your favor. For example, the intelligence community appreciated when groups like ISIS and the Taliban engage in overt online social action in that the activity offered an opportunity to collect information about who they were and what they are up to. Sometimes the enemy can help you win, providing information you can glean from their network. After mapping a network, exploitation can be a precursor to disrupting network activities and actions.

Now it is time to start talking about ways to diminish your adversary. Here is a menu of methods. During World War II, the Allies were so determined to succeed in Normandy that they tried all of them. When it came to disruption, the invasion preparations were the military version of throwing the kitchen sink at the enemy (along with most of the rest of the kitchen).

There are four ways to disrupt a network. Here they are.

Deny. Denying an ability to act is a powerful way to disrupt an enemy. In Normandy, the Allies tried that with the Transportation Plan, denying the Germans the means to get their reinforcements to the Normandy beaches by bombing the heck out of the transportation system. One means of physical denial that is practiced in the cyber world all the time is denial-of-service attacks, employing bots to prevent users from accessing services and websites. This tactic was discussed in chapter 2. There are many ways to organize denial-of-service attacks online, but the most common is to flood a network server with so many requests that the server can't handle all the internet traffic and the flow of information slows down or is interrupted altogether. A distributed denial-of-service attack is when you harness multiple assets to attack a target. This is frequently done with botnets.

There are other ways to deny services. Virtually every online network sits on top of a technology stack or tech stack (also called a digital stack).

The tech stack is the collection of tools, platforms, apps, and software that support the network. This includes everything from cloud services that host databases and operating software to services like PayPal that support online money transfers. Denying access to a component of the stack could well disrupt or shut down network operations. Kara Frederick (mentioned in the previous chapter) has done pioneering research on this threat. Tech stack attacks happen all the time in the real world.[3] That is what occurred to the social networking site Parler. In 2021, over claims the company had violated terms of service, Apple and Google removed the Parler app from their app stores (websites used to download applications on digital devices so that the applications could function on a user's personal device). Amazon Web Services canceled Parler's hosting services contract. Parler was completely disrupted for a month until the company could replace components of the tech stack that allowed for resuming operations. These actions had a dramatic impact. When Parler shut down, it had at least 15 million users and was the most popular app downloaded in the Apple App Store. When Parler came back online, the web service recovered a fraction of its user base and the number downloads at app stores dramatically dropped.[4] By 2023, the number of active users was estimated at about forty thousand.

Ransomware (malicious software that infect a computer and restrict users access until a ransom is paid) is another means to deny critical capabilities to networks. Ransomware has been used by both state and nonstate actors as a means of disrupting governments and private sector entities. As of this writing, the record-setting ransomware strike (that we publicly know about) goes to CWT, a US-based travel services company that shelled out $4.5 million to recover confidential business records. In the process of the attack, the attackers also knocked thirty thousand computers offline.[5]

There are many other tactics that might be employed. This list is long enough, however, to make the point. Find out what the network really needs to operate. Whatever it is, take that down when it is needed most.

Delay. Just slowing an adversary down might be good enough to win the day in the real world or online. Before the Normandy invasion, on June 5, 1944, at 9:15 p.m., the following message was broadcast across the channel: "Wound my heart with a monotonous languor." This was a code, a code for an execution order to the French resistance network, directing that over the next forty-eight hours, they should go on a sabotage orgy. By some estimates, there were at least a thousand attacks (at no small price; estimates are that well over a hundred resistance fighters were killed, wounded, or captured).[6] No Allied commander expected that this valiant sacrifice was going to make much of a difference. There was, however, the hope that it might help slow down the Germans enough to give the Allies an edge for the Normandy landings.

Network attacks that slow performance have the potential for greatest impacts where the exchange of information is most time sensitive and crucial to the network being attacked. Text messaging platforms like WhatsApp, WeChat, Yik Yak, and Telegram Messenger are good examples. These social media apps are popular because they deliver immediate targeted communications. In a national security context, such communications could be vital, particularly during emergency or disaster situations. For instance, during the withdrawal from Afghanistan, the US government set up an evacuation hub at the American air base in Qatar. At the height of operations, they processed and moved over fifty thousand people in and out of Qatar. Initially, the hub was managed by an overwhelmed team from the US embassy, who were later replaced by a staff of volunteers from government agencies. In a conversation with a member of the team, I was told when the team arrived, they had zero infrastructure to manage the system they were put in charge of. The entire operation was managed through a handful of borrowed government digital phones. They shipped fifty thousand people on WhatsApp. Delaying SMS messaging would have completely crippled the operation.

Delaying SMS messaging is one of many malicious activities that can be conducted via malware, software that can manipulate the operation of digital devices. In 2011, a team of researchers built a framework to assess how a significant SMS attack might work. They concluded, "The threat is serious since the attacks can be used to prohibit communication on a large scale and can be carried out from anywhere in the world."[7] Since then, there have been numerous security upgrades to text servicing platforms; nevertheless, attacking SMS apps remains a favored target for cyber bad boys.

Delaying networks at a decisive time and place could significantly impact network performance, cascading into serious consequences that might even cause mission failure. When timely transfer of information is crucial, delay could be a network's most fearsome adversary.

Distract. An enemy distracted can be an enemy defeated. Seb Gorka was a victim of distraction. The media went after Seb while he served in the White House not because he did anything wrong but because accusatory news was a distraction, forcing the administration to react to controversy rather than the president's agenda.

Going back to our Normandy example, the Allies practiced a tactic known as deception. The US military loves deception. They have whole manuals on the subject.[8] Deception operations for the invasion were some of the most elaborate and effective in military history. The guys who pulled this off were literally the mad scientists of misdirection. They made balloons that looked like tanks, simulated radio traffic, and crafted fake amphibious landing craft out of scaffolding tube, wood, canvas, and empty forty-gallon barrels.[9] Based on Ultra intelligence, the Allies knew that the Germans believed these counterfeit forces to be the real thing, staging in Great Britain for an invasion of the continent. It was a total fake out that really distracted Hitler.

The key to the most effective distraction operations is to help convince an adversary to believe or see things in the manner they are predisposed to believe and act on, so you can then go do something unexpected that will totally ruin their day. This is exactly the key plot point in *Ocean's Eleven*. The gang detonates a device they call the "pinch," which generates an electromagnetic shock disrupting the casino security system. After Danny's gang interrupts the casino's surveillance system, they replace the real-time video feed of the vault holding all the cash with fake footage. Casino security is distracted by seeing what they were believing, rather than what was really happening. They completely miss the real robbery.

How to distract a network is limited only by imagination. The key is to create a big enough distraction for the network to notice, accept as credible, and act on (which then gives you an opportunity to exploit their action or inaction in a meaningful way).

Online social media is particularly susceptible to distraction since, by its nature, social media can be, well, distracting. Users are attracted to posts with large numbers of likes or views, which may not at all be significant or useful. Users get overwhelmed with information and can't determine the meaningful from the wrong or irrelevant. Users get seduced by clickbait. By its nature, social media can be used to steer a crowd—sometimes in the wrong direction and to wrong behaviors. One of the most common causes of accidents, for example, is drivers distracted by social media posts.[10]

Distractions on the internet can be powerful and influential. In 2021, a journalist published a picture of a US Border Patrol agent on horseback attempting to impede the entry of illegal immigrants across the border with Mexico. A claim was published on the internet that the agent was whipping the border crossers. The story went viral and almost instantly became dominating news. Even though there were doubts quickly raised about the interpretations of the photos and the claims of whipping were eventually debunked, nonetheless the incident showed the immense power of online social platforms to distract.[11]

Online distraction is a common tactic used by government and nongovernment entities. The Chinese government, for instance, creates content to purposefully distract and influence online communities. One 2017 research study concluded, "We estimate that the government fabricates and posts about 448 million social media comments a year. . . . The goal of this massive secretive operation is . . . to distract the public and change the subject [away from controversial issues], as most of these posts involve cheerleading for China, the revolutionary history of the Communist Party, or other symbols of the regime."[12] The regime considers distraction a principal instrument of social control.

Again, distraction is most effective when it is done for a purpose that furthers your objective, whether it is distracting Hitler from launching his armored forces against an amphibious invasion or the Chinese government exercising authoritarian control online.

Degrade. One way to disrupt an enemy is to take out enough of their capability, which might not be enough to prevent their system from operating but enough to reduce the systems' effectiveness so that it cannot accomplish critical tasks at the critical time and place. This level of degradation could occur through physical destruction or by other means.

On the eve of the Normandy invasion, the Allies knew there were already a bunch of German troops in Normandy. In 1942, Hitler ordered the construction of an Atlantic Wall (*Atlantikwall*), a series of coastal defenses that would be used to repel an Allied invasion. Most of the troops guarding the front line of Hitler's Fortress Europe (*Festung Europa*) were not actually on the coast but billeted in the surrounding areas, ready to rush to beaches when the Allies attacked.

To slow down reinforcing the beach defenses, the Allies conducted airborne and glider landings on the night before D-Day. These troops were never expected to be more than a speed bump to a determined German counterattack. Airborne forces were lightly equipped with few supplies and limited ammunition. If they were attacked by German tanks, mostly all they could do was assault the armor with salty language. The airborne troops were expected to fight the enemy and degrade the German ranks so that there were less enemy to fight on the beaches. This was not intended to be a suicide mission but the nearest thing to it. That any survived to be boot camp instructors in the next war for people like Bob Goldfarb was close to a minor military miracle.[13]

Degrading capabilities is one very visible means of disrupting networks. September 11 offers a harrowing example. The New York City command post for emergencies was in the World Trade Center. That didn't work out so well when the towers collapsed. Likewise, the radio repeaters that were installed to help police and fire communicate inside the building were destroyed by the fire. As a result, many responders were not alerted to withdraw from the buildings before the collapse and died.[14] Degradation of communications significantly hampered the emergency response system that responded to the disaster scene. Those failures in networking cost lives.

Degradation can be a tactic employed in the virtual world as well as against real-world systems and networks. Botnets, for example, can be a powerful tool for degrading cellular networks.[15] Networks where the real-time exchange of information is vital for mission accomplishments are particularly vulnerable to the debilitating impact of degrading system performance.

Offense and Defense

Whether going on offense to disrupt a network or fending off network attacks, it is crucial to understand how systems can be attacked. This introduces the concepts of measures and countermeasures. Measures are the capabilities to conduct attacks. Countermeasures are how to counter the measures. Both of these concepts have utility in planning offensive and defensive operations.

Measures. A measure consists of ways and means. Ways are how to attack. Means are what to attack with. For instance, in the case of Normandy, one way used to disrupt the German defenses was deception. The means for conducting deception operations (described above) was the 23rd Headquarters Special Troop, a 1,100-strong force of artists, advertisers, actors, and engineers who devised and implemented all the ingenious ways of putting one over on the Germans.[16]

Implementing offensive network warfare requires both the ways and means necessary to undertake the chosen method of attack. What is needed depends on what is needed to be done. Conversely, if limited means and resources are available, measures must be scoped to what is practical and achievable with what is at hand.

Scoping measures to the mission is important. China and Russia literally have vast cyber armies at their disposal that can be brought to the fight. More on these later. Others may not have those kinds of capabilities. Actions must be tailored to match abilities. No matter if the assets at hand are ample or meager, if a disruption operation cannot be mounted that can have an influence on an opponent which will achieve the desired impact at the critical time and place, there is a real question if that is the right tactic to fight a networking war. Doing something just for the sake of doing something (such as the hapless American artillery tactics in Vietnam) is never a good idea. Sure, doing "something" might succeed. But then again, it might not. When there are issues of national security on the line, is it worth the risk of getting it wrong?

It is important to be realistic and practical in planning offensive measures. Again, this is where an IPB-like process can be helpful. The process forces you to look at the things that really impact action in the real and virtual worlds, rather than drifting into a dreamscape of wishful thinking that, well, this just might work.

Countermeasures. Countermeasures are, not surprisingly, ways and means to counter measures. There are different kinds of countermeasures. Deciding which are best starts with making realistic risk assessments, a topic covered previously. A measure of risk looks at threat, vulnerability, and criticality and then decides what is the best way to mitigate or counter the risk. Based on a good risk assessment, you can start asking the right question. What are the most appropriate countermeasures to make what my network is doing less risky?

It was, in fact, a bad risk assessment that made Hitler vulnerable to the Allied deception plan. Hitler thought the greatest risk to blocking an Allied invasion was Allied airpower bombing the armored units he had husbanded for the counterattack against an expected invasion. To mitigate the risk, he disbursed his armor reserves so that they would be harder to be found and attacked before the Allies hit the beach. Hitler solved one problem, but he created another vulnerability he did not adequately address—his force was vulnerable to disruption when ordered to deploy from assembly areas to the beaches.

Planning countermeasures should be axiomatically an important part of protecting national security networks from debilitating disruptions. Even without an enemy coming after you, there will likely always be what Carl von Clausewitz called the "fog of war" or the "friction of battle." Bad stuff happens. Servers crash. Software gets bugs (this problem was literally named after a moth that fouled up one of the first computers). Yahoos at work click on innocent-looking websites and download malware. Such is life.

Smart, serious people plan on everything going wrong. That's what Dwight Eisenhower and Marshall did (a big part of the reason why we won World War II). In our equally competitive world of social networking warfare, planning for resilience by having suitable countermeasures in place is an important part of staying online, or at least staying alive.

Countermeasures might include protecting systems against attack, having redundant systems, having alternative means for operating, or ways to quickly repair or replace damaged equipment or systems. Many networking platforms have plans to either ensure business continuity, the ability to keep operating in the event of a threat to the network, or emergency response plans to recover capabilities if lost at a critical time.[17] For instance, a resilient platform will have one or more contingencies to ensure operations if a critical part of their tech stack is unavailable. There are, for example, already companies offering tech stack services that providers claim can be used without "fear of being monitored, manacled, censored, or deplatformed."[18] Social media messaging can also be a tool to help ensure operational resilience, providing a means for sharing vital information.[19]

What is important to remember is that measures and countermeasures, particularly in the social networking world, are not always physical actions or cyberattacks. They can be psychological as well. The disruption that led to Gorka's fall and rise was largely structured around issues of reputation and credibility. They created questions and doubts that made deciding and acting for others more difficult. Indeed, the reason for Seb's powerful presence in the social media world today is his uncanny ability to zero in on finding issues that question the trustworthiness of those who claim they are looking out for America's national security. Are they really?

War Face

This chapter explained how to annoy your adversary enough to outcompete their network. The chapter covered four different methods of disruption—deny, delay, distract, and degrade. Any one of them, or in combination, might be enough to ensure you win and they lose. If a network warrior puts together a better combination of measures and countermeasures, they will likely come out on top. But maybe not. Disrupters don't always win.

Some adversarial networks are tough and resilient and may not make stupid mistakes like Hitler. (Hitler was happy the morning he found out about the Normandy invasion because he assumed his armored counterattack would crush the enemy on the beachhead. Defeated and humiliated, his allies would surely sue for peace.) For determined and less stupid enemies, winning might require something more potent. Winning could mean having to decisively defeat the other side. That is what the next chapter is about.

9

BAD FOR BREAKFAST

Nadia Schadlow is the smartest person I know. She hates it when I say that. It might, however, be true. Here is why.

In the tumultuous first year in the White House after the tumultuous presidential election of 2016, the tumult was nowhere more tumultuous than in the big hulking gray building next to the White House on Pennsylvania Avenue.

The official name of this structure is the Eisenhower building named after the general and president. Most people, however, just call it the Old Executive Office Building for the obvious reason that there is a newer executive office building. The "old" office building was built in the late nineteenth century to house the State, War, and Navy Departments. Considering the size of the State and Defense Department today (the Pentagon, built at the direction of George C. Marshall by the way, alone is the largest office building in the world), shows you how much government has grown since. When those folks moved out, the expanding office of the president moved in. Those not lucky enough to score an office in the White House West Wing wind up in the echoing vaulted halls of the Eisenhower Executive Office Building (which looks a lot like the Haunted House ride at Disneyworld), including the staff of the National Security Council, the body created after World War II to help the president manage foreign and defense affairs and coordinate between the different federal agencies, like the State and Defense Departments.

All this history matters because who runs the National Security Council staff matters. It is part of their job to help to explain to the rest of the government, Congress, the American people, our allies, and our adversaries what is the foreign and security policy of the United States. This is principally done through the publication of the National Security Strategy, a document that the president is required, by law, to publish explaining exactly what the strategy is that guides US policies and actions. Anyway, that is what the document is supposed to do. Often, national security strategies are pretty much a nothing burger, meaning they don't amount to much, because they literally are less a strategy than a long, windy glossy book report that just lists what government agencies are already doing or wish would happen, rather an actual statement of a determined action-oriented strategy directing how the United States was going to engage the world. There are exceptions. Dwight Eisenhower, for one, as president was serious about strategy. Many others not so much.

Strategy was important for this president because most people had no clue how the president planned to tackle America's problems. The most concrete statement made during the campaign was "America First" (discussed in a previous chapter). That declaration did not explain much. A huge controversy erupted over what the bumper sticker even meant. Some claimed the president planned to be an isolationist (the strategy the America First committee argued for before Pearl Harbor). Others argued no, it just means looking after American interests first. Even that answer wasn't very satisfying. Not everyone agreed on what were America's interests. And, of course, the campaign slogan didn't address other important questions: What about everybody else's interests? What about our allies? What about our adversaries? What did the president think about them? Inquiring minds wanted to know.

This was not automatically a cause for panic. Presidents say all kinds of stuff on the campaign trail. Woodrow Wilson and Theodore Roosevelt promised to keep us out of world wars. Richard Nixon said he had a secret plan to end the war in Vietnam. On campaign, presidents say stuff to get elected. It is what they do as president that matters in national security and foreign policy. You don't start keeping score until the candidate raises their right hand on the steps of the Capitol, and then, well, that is what a president has a staff for—not to think for them but to shape and explain what they really want to do in documents like the national security strategy. For this president, however, that turned out to be a real problem.

It all started out well enough. The president appointed a national security adviser and a deputy. They oversee the council staff. So far so good. They reported for duty on day one. As also did other national security policy advisers like Sebastian Gorka. That was, however, when things started to, as they say, go off the rails. Let me describe the day I knew that national security was in for a raucous roller-coaster ride.

The "passing of the baton" has become one of the traditions of Washington presidential transitions. This conference, hosted by the US Institute of Peace, brings together members of the incoming and outgoing National Security Council staffs, as well as national security luminaries from left and right, from the think world and previous administrations to share a moment of a dialogue and discourse on the importance of bipartisanship and how to serve the United States' interests.

I remember the 2017 event well. At the last minute, I was drafted to help organize a bit of it. Honestly, they must have expected the other candidate to win. They were well prepared to facilitate that transition since the incoming and outgoing people would have been pretty much the same people. When the expected didn't happen, the organizers scrambled to find anyone who knew anyone who knew anyone in or anything about the incoming team. I was a convenient target; not only had I worked on the transition, but many of the folks on the national security team, including the national security adviser and his deputy, as well as Gorka and others I had known for years—one of the by-products of networking.

The event started fine enough—receptions, cocktails, dinner. Then it got very awkward, very fast. The day that the deputy national security adviser spoke was also the day news about the infamous Steele dossier, accusing the president of colluding with the Russians, broke in the press.[1] How we found out about it was when the deputy national security adviser finished her remarks. The first question she was asked was from a reporter who said something like, "What comment do you have about the dossier accusing the president of colluding with the Russians?" The deputy national security adviser, like most of us, had no idea what the reporter was talking about. It was pretty much, as I remember, the first many had heard about the breaking story.

This incident is worth citing here because it marked a tumultuous turn of events that triggered an open season on many presidential advisers. The national security adviser resigned. The deputy resigned because she assumed his replacement would want to pick their own deputy. Others followed later like Seb Gorka (who was not on the council staff and left for reasons not associated with the Russian collusion allegations). This left a Dyson-scale power vacuum in the White House national security–making process.

That void was filled by two people. One was an old army acquaintance (more networking); H. R. McMaster, a three-star general, was asked to serve as national security adviser. Before his appointment, H. R. called me and asked me if he should take the job. I enthusiastically said yes. I still think I made the right call and will explain why later. General McMaster made some crucial and important contributions during his tenure as national security adviser.

This is where Nadia Schadlow comes in. McMaster asked her to serve as his deputy. And this is where our story picks up.

I have known Nadia for a long time. For many years, she was the senior program officer in the International Security and Foreign Policy Program at the Smith Richardson Foundation. It was a perfect position for her. This job allowed her time to raise a family (outside the Washington bubble in a community full of normal people), and work on a PhD supervising grants at Smith Richardson was also super interesting and impactful work. The foundation provides financial support to scholars and organizations in the national security field, one of the few major private grant-making institutions that really invested in building up the United States' arsenal of national security brainpower after the Cold War, when most others assumed the end of history was at hand and national security would not be much of a challenge.

Nadia got her PhD from the Johns Hopkins Nitze School of Advanced International Studies, becoming an established scholar in her own right. She served on the Defense Policy Board from 2006 to 2009 with folks like some guy named Henry Kissinger.

I won't say she is a prolific writer (I hounded her for years to turn her dissertation into a book, which eventually she did in 2017: *War and the Art of Governance: Consolidating Combat Success into Political Victory*), but her scholarship is careful and articulate. When she does write in places like *Parameters*, the *American Interest*, and the *Wall Street Journal*, it takes reading only a couple of sentences to figure out. I was not exaggerating much: She is brilliant. That's why whenever we did public events together, I would always start by saying, "Nadia is the smartest person I know." I knew it embarrassed her, but hey, people needed to know (even Albert Einstein could have used a good publicist).

It is little secret that Nadia was the principal author of the 2017 National Security Strategy, though like most government documents, lots of people had a hand in a sentence or a paragraph or two here and there.[2] All that said, here is what Nadia will tell you: The real author of the document was the president. What she and the National Security Council staff did was listen to the president and put his vision for American freedom, prosperity, and security into the framework of a national security strategy that articulated ends, ways, and means.

The core of the strategy focused on what the administration saw as the key threat to US power—great power competition, the countries that had the means and intention to create problems for the world. The president's strategy articulated how the United States would defeat them. The United States didn't need to eliminate its rivals. What the United States intended was to defeat their actions, bending their will to ours and making them give up on trying to diminish America. This strategy described well how the president conducted foreign policy (as opposed to what all the punditry, politicians, and press described as the foreign policy). Furthermore, I would argue that it was what our allies and adversaries thought our foreign policy really was—which, by the way, is the purpose of what these documents should for be in the first place.[3]

Whether folks like or hate that president and his policies is not relevant here. The main point is that his administration did what a president is supposed to do in matters of national security. If winning is important, they ought to have a suitable, feasible, acceptable plan to do that and then get about doing it. That was what they did.

H. R. deserves this country's gratitude. First, he hired Nadia Schadlow. Second, he brought rigor and structure to the national security–making process when it needed it most (much in the way that Kissinger did for Nixon, Richard Allen did for Ronald Reagan, and Eisenhower did for himself). The framework his team laid out served throughout the presidential term.[4] Together, H. R. and Nadia offer an object lesson in understanding the concepts of winning and losing in national security competition and how the concept of defeat fits in. This is a lesson that has application to social networking warfare.

From Adrift to a Drift

A previous chapter harped on the importance of having a plan. What George Marshall thought of as the "guiding lifeline of an idea." One of the most vital elements—okay, actually the most vital element of the plan—is the "mission." What are you trying to accomplish: storm the beaches of Normandy, take down ISIS, or silence a critic online? This cluster of chapters examines how to go after a network by going after the network's mission. What do you need to do to your adversary's network to win? This chapter covers three options—disrupt, defeat, and destroy. The right answer might be one or more, in sequence or en masse. This chapter covers defeat.

A previous chapter that reviewed designing tactics described Carl von Clausewitz's concept of the culminating point or overreach. Fighting for national security is not for the timid. Risking too little will likely mean you will always lose, like the poker player who folds when anyone bets over a dollar.

In contrast, throwing too much into the fight, particularly at the wrong places and time, might make you just as vulnerable to defeat, like gambling it all when you have a pair of twos.

Disruption, described in the last chapter, might often be the least demanding task when it comes to putting skin in the game to outplay your adversary. When national security is on the line, however, more commitment may be required. Sometimes, disruption is only useful as a prelude to something bigger, as was illustrated in the example of the Normandy campaign. When more is what is called for, the next step up the ladder of escalation is defeat.

Before attempting to plan the defeat of an enemy, two questions must be answered. First question: it is not enough to ask, "How do you defeat your competitor?" The question that must be asked first is, "What is defeat?"

A previous chapter describing tactics reminds that national security competitions are not really games with fixed rules. In most games, it is easy to know who is defeated. The rules explain it. Who scores the most points? Who gets tagged? Whose Marco gets polo-ed? National security competition does not have fixed rules.

Defeat is also a complicated concept in part because it is not always the same thing. Defeat can be as much psychological as physical destruction. Clausewitz once wrote, "Defeat is in the mind of the enemy commander." In other words, no matter how much punishment is inflicted, an enemy is not defeated until they give up. Just ask *Rocky* (1979).

There is a movie that illustrates the concept of defeat fabulously well, even more so because it is based on real-life events. In 1879, the Zulu tribe decided they would take on the British army. Indeed, they massacred a good portion of the British expeditionary force at the Battle of Isandlwana (described in *Wiki at War*). Part of the Zulu army, some 4,000 warriors, then marched on the small garrison at Rorke's Drift (a river crossing station) manned by about 150 soldiers, commanded by two lieutenants, John Chard and Gonville Bromhead. In the film, after they learn that the main force defending the area had been wiped out, they decide the only way to defeat the Zulus is to hold out until the main British army can counterattack. In the film, the conversation goes something like this:

> Lieutenant Chard: "The army doesn't like more than one disaster in a day."
> Lieutenant Bromhead: "Looks bad in the newspapers and upsets civilians at their breakfast."

They decide to defend the station. After two days of fierce fighting, the Zulus give up and leave—defeated. If you want to know how, well, that is another story.[5] The important lesson here is that winning starts by understanding losing. The alternative is bad for breakfast.

Definition of Defeat

The challenge of understanding defeat reminds us of the adage, "The enemy gets a vote." The objective of a competition is to impose your will on the adversary. The adversary, however, gets a say when enough is enough. The wars in Afghanistan and Vietnam are good examples. Despite the brutal and relentless punishment heaped on them, the bad guys won, simply because they refused to ever lose.

There are a few more illustrative examples to explain the illusive concept of defeat than the two wars the United States fought in Iraq. In the first Gulf War (1991), the United States believed they had won. They liberated Kuwait and rolled up the Iraqi army in a campaign that could be measured on a watch, not a calendar. After the victory, Chairman of the Joint Chiefs of Staff General Colin Powell even promulgated a doctrine of "decisive victory," which in turn also expressed a concept of "decisive defeat." For Powell, the swift and humiliating war against the Iraqi army was the penultimate example of decisive defeat. Powell's nemesis, the Iraqi leader Saddam Hussein, however, didn't see it that way. He rebuilt his army and continued to threaten US interests. In the end (rightly or wrongly), the United States, with Powell now serving as secretary of state, felt compelled that defeat required ousting Saddam, abolishing his army, and occupying the country. That required a protracted campaign (2003) that in the end looked nothing like the Powell doctrine.[6]

Don't pick on Powell. Many great commanders in history failed because they did not get defeat right. Napoleon thought the Russians would be defeated when he beat their army (which he did at the Battle of Borodino, September 7, 1812) and occupied the Russian capital. Wrong. Douglas MacArthur thought he would defeat the North Koreans when his forces reached the Yalu River (October 1950). Wrong. History reminds you that you can get a lot about competition right, but if you get this part wrong, the one defeated could be you.

The concept of defeat has relevance to networking warfare in the real and digital worlds. Take the example of Russian malicious networking efforts to destabilize and spread misinformation in Europe and the United States. All the Western governments complain about Vladimir Putin's meddling. They impose sanctions. Issue indictments. Seize assets. Toss in some cyber countermeasures. Even when his efforts are successfully bested (like when my friends in a small central European country outwitted him), Putin, like

The Terminator (1984), keeps saying, "I'll be back." If these countries really wanted to stop him, they would have to define a goal to defeat his cyber networking weapons. They would have to commit to defeating a brutal, ruthless authoritarian ruler with considerable resources who is answerable to no one. There will be a longer discussion about state adversaries like in Russia in following chapter. But for now, let's just say, that's not happening anytime soon.

Defining Defeat

In 2006, Jan Angstrom and Isabelle Duyvesteyn authored an interesting volume of collected essays, *Understanding Victory and Defeat in Contemporary War* (2006). In summary, the authors pretty much agree with Clausewitz: Victory and defeat are transactional choices. Just like leadership is a relationship between the leader and the led, winning and losing are implicitly or explicitly a relationship acknowledged by the victor and the vanquished. To be clear, this is not about reaching a common understanding. Enemies rarely do. Defeat occurs when both sides reach a common perception on the resolution of the competition.

Sometimes, defeat is readily apparent by joint agreement. The Japanese, for example, acknowledged they were defeated when they signed the surrender document on the deck of the battleship USS *Missouri* (September 2, 1945).

Other times, defeat is de facto. Although the Germans signed a surrender document on May 7, 1945, they intended to keep on fighting just like the Taliban after 9/11. There was, in fact, an elaborate insurgent warfare plan and organization, SS Werewolf, that was supposed to carry on the German struggle after the army surrendered. The Americans expected fighting would continue. They even feared the Germans had a secret Alpine Redoubt where the last fanatics would hold out. The Nazi will to resist, however, quickly crumbled. It turned out that the Germans did think Germany was defeated.[7]

So how do competitors decide when to call it quits? Understanding the relationship between victor and vanquished requires returning to the discussion of interests. Competitors use their level of interests to determine how much they are willing to fight. The more important the interest, the more the competitor is willing to compete to secure that interest. When the interest is vital, competitors may well be willing to invest an extraordinary effort if the competitor believes there is still a prospect that those interests can be secured. For instance, even though the United States had far more resources than North Vietnam, conquering the South was a much greater vital interest to the North than protecting South Vietnam was to the United States. They were willing to expend more than we were. We gave up first.

So, defining defeat requires understanding your level of interest and the resources you are willing to expend, as well as understanding your competitor's level of interest and the resources they are willing to throw into the fight. Defeat is when one side recognizes the balance is irrevocably not in their favor and there is nothing they can do about it.

It was impressive to watch Nadia subtly weave into the national security strategy a concept actually rarely seen in post–Cold War US strategies, an actual theory of victory, an underlying idea of how the United States could achieve the conditions of "we win, they lose." It was more brilliant in that it eschewed maximalist solutions like regime change, nation building, and threatening war. Nadia crafted a strategy that leveraged US power against our adversaries' weaknesses. Her plan was an exemplar of good strategy in part because it responded to the reality of contemporary geopolitics, taking the capabilities and intentions of all sides of the competition into account.

In crafting a concept for the enemy's defeat, the calculations on both sides can be dynamic. Before and during the conflict, competitors may well change their mind on how they see things. Americans didn't think World War II was worth fighting. The day after Pearl Harbor, they changed their minds. During a competition, missions might change or even be abandoned. Adolf Hitler, for example, saw Britain's defeat in the invasion and occupation of the British homeland. He gave up after failing to win the Battle of Britain, the German air offensive against the United Kingdom (July–October 1940). Defeated, he came up with a new mission, defending Fortress Europe at the Atlantic Wall. That, of course, did not work out so well either. Like every aspect of competition, defeat can be a dynamic concept.

Competitors can change their mind with good reason. Defeat is just a planning tool, not a guarantee like a warranty on a toaster. In competitions, both sides act and counteract and must adjust to the reality of shifts in the competition, particularly when seeing an opportunity to win or faced with the harrowing prospects of losing. That said, there is not just "no substitute for victory" (MacArthur's famous saying); there is no substitute for having a concept for what constitutes a real defeat. A clear definition of defeat is essential for planning serious action-oriented efforts.

Here is a good example. When the United States entered World War II, country's leaders did not make the decision to conceptualize defeat as the total surrender of the enemy's armed forces lightly. The US government assessed that their enemies would fight resolutely and would have to be conquered with overwhelming force. The United States entered the war believing it could defeat its enemies. Even before the war began, the US military concluded that America (in concert with allies) could muster the means to win.[8] Without that confidence, they would have never been able to craft a plan to defeat Germany and Japan.[9]

The dynamics of defining defeat relate to networks as well. Take the case of the social networking platform Parler (discussed in the last chapter). There is no question the platform's detractors managed to disrupt the operations of Parler. But was it defeated? The backers of Parler proved resolute in their conviction that other big tech social networking platforms were a threat to free speech and all too willing to censor opposing political views. Parler's backers also had deep pockets. They were ready to invest in rebuilding Parler's tech stack and getting the platform back online. As of this writing, they have demonstrated their level of interest and assets outweigh those of the app's enemies. Parler may be down, but it is not out. The platform got back online. It was disrupted and definitely diminished, but was it defeated?

Even in distributed networks (where there may not be a leader or funder who can decide for the network it is time to surrender), the concept of defeat has relevance. On the one hand, online social networks can be very resilient because they are not centrally managed and controlled. There is no single target to decisively influence. On the other hand, distributed networks can also be extremely fragile. Their strength comes from the accumulation of users. When users abandon the network, or what connects the network is removed, the network collapses.

Targeting users in a distributed network can precipitate defeating the network. Chapters 3 and 4 discussed the importance of recruiting and retaining members in an action-oriented network. Recruiting requires a credible, understandable, and actionable message. Retention requires an active, effective value proposition. If a competitor can undermine the believability of any of the value propositions, a network might collapse because the members of the network abandon it. Another concern is the threat of social action, such as targeted consumer boycotts that attack a company or organization for actions or beliefs.[10] This is one reason why many online platforms and social media influencers invest heavily in defending their reputation online; they know compromising their credibility with customers or followers could be a death blow.

Defining defeat is important for networking warfare as it is for real war. But how do you define defeat? The answer is the same online and offline: It depends. Since defeat is essentially the competitor accepting that their army or network can longer accomplish its mission, defining defeat is best expressed in terms of preventing mission accomplishment. If they don't have what it takes to accomplish the mission, they lose. A definition of defeat might address defeating capabilities (e.g., the capacity to defend the homeland), time (e.g., preventing accomplishing a task by a certain date), geography (e.g., preventing crossing a river), or task (e.g., stymieing organizing a presidential transition team). It depends. The right expression of defeat is that when you take away what your competitors believe they must be able to do to win—whether that is assemble adoring fans or armies and air forces—they lose.

Designing Defeat

Answering the question of what constitutes defeat raises the next question—How do you defeat an enemy? One way is to try something. Then get a time machine and race to the future. See if your plan worked. If the plan did not work, race back to the past and keep shuttling back and forth until you get it right. Absent a time machine, what planners do is construct a "theory of victory," envisioning a suitable, feasible, and acceptable way to defeat their enemy.[11] This conceptualization includes an expression of ends, ways, and means. Envisioning a future correctly is as close to time travel as planners can get.

The above discussion helps answer parts of this question—thinking through what constitutes defeat of your enemy and if you have the capacity and will to carry through taking them down. If these issues are thoughtfully and appropriately addressed, they help formulate the ends (goals) and the means (resources). How to defeat the enemy, however, must also explain the "way," the actions by which you will defeat the enemy.

There are a number of ways that might suit for defeating an enemy in networking warfare. Which one suits best? Well, again, that depends on both what you and your competitor bring to the fight. This reminds again why the Intelligence Preparation of the Battlefield process described in chapter 7 is useful. That framework can help refine which method of defeating your adversary makes the most sense. To help, here is a short list of conceptual planning tools for thinking through how to take down the other side.

Direct Approach. One option is a direct assault (this concept was introduced in designing tactics in chapter 3)—go straight at your enemy and try to defeat them. Clausewitz is considered a strong proponent of the direct approach. Identify your enemy's strength and then go crush it. Sounds simple, direct, and powerful. It is. It also doesn't always work. The direct approach was Hitler's plan for the Battle of Britain. He failed. The direct approach was the Allied idea for the Normandy invasion. That worked okay.

In networking warfare, a direct approach would mean directly attacking the structure, users, or the activities of the network.

Indirect Approach. Also discussed before, this option seeks to avoid attacking the competitor directly and instead focusing on a weakness or vulnerability to facilitate defeat. This approach was popularized by the British strategist B. H. Liddell Hart in his book *Strategy* (1960). MacArthur used the indirect approach in his island-hopping campaign during World War II, bypassing enemy strongholds.[12] This strategy works in the network world as well. For instance, the assault on Parler did not take on its network of fifteen million users directly; rather, it undercut the platform by eliminating the tech stack that allowed the app to operate.

Attrition. This is another way to think about defeating a competitor—just keep whittling down their capability until they can't fight anymore. In his book *The American Way of War: A History of United States Military Strategy and Policy* (1977), the historian Russell Weigley argues this is how the US military usually fights. There is no question some US campaigns were planned that way including the American Meuse-Argonne Offensive (1918), of which one of the principal planners of the campaign was George Marshall. In the online space, botnets are a good example of attrition warfare. They just keep assaulting until they break down the network's defenses and they get what their botmasters want.

Maneuver. This concept was introduced as an alternative to attrition warfare. Rather than just crushing stuff, campaigns are organized to surprise the enemy by attacking in unexpected places that unhinge the enemy. This approach was popularized by William S. Lind in his treatise *The Maneuver Warfare Handbook* (1995). One often cited example of maneuver warfare is the German assault through the Ardennes (1940), which surprised the French army and precipitated the rapid defeat of France. Spoofing, spearfishing, and similar malicious cyber techniques are arguably forms of digital maneuver warfare. Rather than trying to bust down cyber defenses like bots battering a firewall, they look for ways to bypass security measures posing as legitimate users to get inside and then attack a network.

Cost Imposition. This approach pushes an enemy to lose by overinvesting in trying to win (violating Clausewitz's culminating point principle). In other words, it entails making the cost of fighting or defending so intolerably high that the other side gives up or collapses. Paul Kennedy described this strategy in *The Rise and Fall of Great Powers* (1987). In part, the Soviet backed both the wars in Korea and Vietnam as well as insurgencies in Latin America and Africa and transnational terrorists in the Middle East in hopes that this would contribute to overwhelming the West. In the online world, one of the criticisms of big tech is that they often crush smaller competitors who cannot match the vast resources of dominating social media platforms (more on this debate to follow).[13] Indeed, companies like Google often wage lawfare, going through costly litigation because they know their competitors don't have the time and deep pockets to pay for lawyers.

Deter/Dissuade. On occasion, enemies can be defeated by being convinced not to act to begin with. This was a fundamental premise of the US strategy of containment during the Cold War, described in *Strategies of Containment: A Critical Appraisal of American National Security Policy During the Cold War* (2005) by the scholar John Lewis Gaddis. Deterrence was a critical element in the US effort to forestall a Soviet nuclear attack. In the online world today, the United States relies heavily on the cyber capabilities of the National Security Agency to deter enemies from attempting something like a cyber–Pearl Harbor.

The Essence of Defeat

This chapter ought to give an appreciation for the seriousness and complexity of the effort H. R. and Nadia faced in crafting the president's national security challenge, as well the difficulty of the task any planner faces in developing a plan to defeat a determined and capable foe in the national security space. It is no mean feat. To win much, much may have to be risked.

Offense and Defense

The concepts described above can be used to organize a campaign of offensive and defensive actions aimed at defeating a competitor in a socially networked world.

An illustrative example of this type of warfare in action comes from an Australian researcher who examines the online war over the debate of COVID-19 vaccinations.[14] The researcher concludes that as the online presence of organizations become more important in the debate, the more they became targets for concerted attack, with the attackers seeking to identify and exploit vulnerabilities to either deter the online participation of users or undermine the organization's ability to operate.

What the Australian study describes includes all the attributes of classic offensive and defensive action (as well as many of the ways and means discussed in *Digital Dominance*). The research determined, "the main modes of attack are disrupting discussions, dominating and descriptions and ridiculing and intimidating opponents. The main modes of defence are excluding disrupters, providing counter-descriptions, making formal complaints, and ignoring or exposing abuse." This laundry list of fighting techniques illustrates the concepts of national security competition are as relevant in the online world as they are in the real one.

Technology

Technology can be a major factor in imposing or staving off defeat insofar as it provides to enable offensive operations or reducing vulnerabilities that could make a competitor more resilient against attack.

Cybersecurity tools (as noted in a previous chapter) are, without question, important assets in networking warfare online. They are, however, not the only technological tools that might be of significant value.

As was noted, one powerful line of attack against networks is reputational threats. Automated means to detect and counter reputational attacks are becoming increasingly common. They will likely be seen as even more important in the future.[15] Deep fakes, discussed before, are also being used more frequently to attack the credibility of networks and networkers. New technologies are also being deployed to help address them. For instance, Microsoft is developing a technology to provide assured voice authentication.[16]

Technologies are only going to heat up the competition in the years ahead. Emerging technologies, which will be described in chapter 14, are going to deliver a powerful new suite of measures and countermeasures for network fighting. If a competitor is serious about the defeat business, they are going to spend a good deal of time keeping up with the latest developments—or a good deal of time looking over their shoulder.

Going Full Monty

This chapter described another means of dealing with competitors in the national security networking space, covering what defeat is, how to defeat an enemy, and describing a set of tools for planning action against the enemy.

What happens, however, when you face a serious enemy that just won't quit? What happens if you can't figure out a way to disrupt or defeat an enemy and losing to them is just not an option? Well, then the next step might be that you must (like Don Corleone) get rid of them altogether. How do you destroy an enemy's network? This topic is next.

10

CRUSH YOUR ENEMIES

Pete Mansoor was a professional destroyer.

When I thought about writing this chapter, I could think of no one who better exemplified what the concept of destruction is all about.

The statement above was not intended as glib. There is no joy in reveling in the misfortune of others. Suffering and annihilation are curses on humanity. Still, in matters of national security conflict and competition, sometimes justice demands the heaviest of hands. Yet, even when the only option is an overwhelming, devastating response, the arm that wields the death blow ought to be guided by a sense of responsibility, compassion, and integrity. Those character traits define Pete.

Peter Mansoor is a professional military historian. He holds the General Raymond E. Mason Jr. Chair in Military History at the Ohio State University. Before that, however, Pete was a soldier once and young.

Retiring as a colonel, Mansoor spent twenty-six years in uniform. Our paths have crossed over the years. Peter graduated from the military academy at West Point. He taught in the history department at the academy. He authored an impressive award-winning book on combat in World War II, *The GI Offensive in Europe: The Triumph of American Infantry Divisions, 1941–1945* (1999). He commanded combat troops in Iraq. His unit was awarded a Presidential Unit Citation for valor, an experience he captured in another book, *Baghdad at Sunrise: A Brigade Commander's War in Iraq* (2008).

In 2007, President George W. Bush ordered General David Petraeus (whom I taught with at West Point; Dave was in the social sciences department) to overhaul US strategy for dealing with the troubled occupation, leading to the suppression of the uprising in the Sunni Triangle. I attended a briefing at the Pentagon with Petraeus before he left. When it was time to take questions, I asked the first one. I said, "Dave, I am really troubled about what you are doing." He gave me a quizzical look. The room went silent. It was already common knowledge he had asked several veterans of the social science department to accompany him as advisers. I said, "I am worried you are taking too many soc[ial science] guys." He laughed. There was always a gentle rivalry between the social science and history departments at the academy. Then he retorted, "It's okay. I am taking Pete Mansoor as my executive officer." I said, "Well, I guess you will be all right, then."

Off to war they went. From that experience came another book by Pete, *Surge: My Journey with General David Petraeus and the Remaking of the Iraq War* (2013).

In his years of service, Pete, like another one of our West Point history department colleagues H. R. McMaster, earned a well-deserved reputation as a soldier scholar. These officers are special; not only skilled warriors who crushed more than a few enemies, they could also think long and hard about the appropriate, restrained, and responsible use of force to accomplish national security objectives. They could teach about it. They could write about it. While H. R. displayed this skill in overseeing the writing of the 2017 National Security Strategy, Pete does it day in and day out teaching and mentoring students in the practice of writing and researching military history. Few are more accomplished in this endeavor. Pete is an expert in understanding destruction.

Our Violent Past and Present

I admit there is often a light touch in *Digital Dominance*. Serious subjects sometimes are more palatable when taken with a little levity. There is nothing unserious, however, about the concept of destruction. Serious people like Peter Mansoor remind us of that.

Just because this is a serious subject, however, doesn't mean we can't talk about movies.

"What is the best life?" asks a merciless, bloodthirsty warlord in *Conan the Barbarian* (1982), to which Conan replies in his best Arnold Schwarzenegger accent, "To crush your enemies, see them driven before you, and to hear the lamentation of their women!" Can't think of a better explanation of destruction than that.

There are, of course, far more serious ways to learn about the hard part of competition than sword and sorcery films. Real wars, like the ones Pete Mansoor lived through and writes about, are way better exemplars.

Yet, another reason why *Digital Dominance* frequently falls back on military example analogies is not just because they are good archetypes of competition; they also remind network warriors that even though networkers may not be trading bombs and bullets with their adversaries, the consequences of their action may have just as much weight. When we pulled together our Afghanistan response group, for instance, the members knew full well people might die if we screwed stuff up. Our group benefited from folks like Terrell Chandler who had been in real firefights. Still, it doesn't take a combat patch to make for a serious fighter in the networking world. What is required is the mental armor for combat. To build that strength for the fight, military history offers a powerful intellectual tool for thinking about any kind of combat.

Arguably, this kind of hard thinking is in decline, notwithstanding the efforts of scholars like Pete Mansoor.[1] The distinguished war historian Max Hastings laments, "In centers of learning across North America, the study of the past in general, and of wars in particular, is in spectacular eclipse." The serious teaching of military history, he argues, is in trouble. His verdict created a bit of a buzz among national security professionals.

Hastings highlights one reason military history is dying on an academic hill at the foot of the ivory tower. "The revulsion from war history may derive not so much from students' unwillingness to explore the violent past," he suggests, "but from academics' reluctance to teach, or even allow their universities to host, such courses." The presumption in much of modern academia appears to be that (1) only warmongers would teach about war and (2) most military history, like much of history, is a tool of institutional oppression and control. As another historian noted, "Unfortunately, many in the academic community assume that military history is simply about powerful men—mainly white men—fighting each other and/or oppressing vulnerable groups." This is putting political correctness ahead of facts. Practicing national security without understanding military history is like doing surgery without ever having taken a class in anatomy.

The value of military history is less in remembering dates and place names of battlefields than in using history as a tool for sharpening intuitive judgment, one of the key skills, knowledge, and attributes of networking. Violent competition is a place where both the linear abilities of disciplined organized thinkers (like a George Marshall) and the creative nonlinear conceptualizers (like a Napoleon) when combined make for the most formidable fighters on the field.

At its core, military history is about people making hard choices based on what they know about themselves, the enemy, and the place they are fighting and living on (or dying over). History is a laboratory for hard thinking, and your brain is the lab rat. We all know the smarter rat beats the maze and gets the cheese.

If you plan on taking the recommendations of this chapter seriously, then work on your skill set. There is no better place for skilled practitioners to mentally rehearse than in the practice of destroying an opponent. The decision to destroy can be deeply consequential. The actions of following through on crushing an enemy can be even more so.

Destruction 101

There is a big difference between defeat (discussed in the last chapter) and destruction. Defeat means a system or network cannot accomplish its mission. Destruction means it cannot accomplish any mission. The allies, for example, didn't destroy Germany or Japan; both nations were defeated. Then we helped them rebuild and they became cool allies. Destruction is what Rome did to Carthage. See if you can find Carthage on a map. You can't. In 146 BCE, the Romans stormed the city, leveled every building, killed or drove off everyone (population of about half a million), and sold the last fifty thousand into slavery. That's an example of destruction.

Why destroy an enemy? One reason is that when competitors believe their adversary is such a threat to a vital interest, destruction is the only option. This is not just the stuff of biblical tribes and ancient civilizations led by the likes of Roman emperors and Atilla the Hun. In the early days of the Cold War, for instance, before the United States settled on the strategy of containment, nuclear experts and national leaders argued that the Soviet Union represented such a grave threat that the United States should launch a preemptive, preventive atomic attack to destroy the Russians—before they attacked America.[2] Indeed, in modern times people practice destruction all the time through the practice of genocide, such as China's reprehension of Tibetans and Uyghurs (more on this later).

The other reason to destroy an enemy is because it is a more complete way to eliminate a competitor than trying to disrupt or defeat a system or network. Previous chapters, for example, discussed the difficulty of affecting the performance of nonlinear networks. Not knowing how they might respond to external attack or striking at parts of the network, it may be difficult to determine how to achieve desired effects that would ensure the network was not a threat. An alternative is just to destroy the whole network. Problem solved. If the network cannot do anything, it is no longer a threat.

Should an enemy be destroyed? Here again, the Intelligence Preparation of the Battlefield framework has utility. The framework helps thinking through what an enemy might to do and how to counter it. An assessment might conclude that if eliminating the threat is really that important, the only practical answer may just be to go all Conan the Destroyer.

Is there a moral and ethical component to this choice? Yes. This concerns the issue of "proportionality." In just conflict, the act of intentional violence should not be excessive compared to the objective.[3] In fighting in Iraq, for instance, Pete Mansoor faced these kinds of ethical choices all the time. Insurgents would try to blend in among innocent villagers. He constantly had to weigh how much force was appropriate for an operation. Just destroying the whole village was always a disproportional unacceptable choice. (Of course, evildoers don't care about ethics; they will just destroy their enemies and innocents because they can.)

Rightly or wrongly, history reminds there will always be competitors that try to go all the way. Therefore, it is worth thinking about the unthinkable because, trust me, competitors will.

How to Destroy an Enemy?

There are two kinds of destruction. I literally made these categories up because I could not find any other descriptors that served sufficiently well. One category is wholesale destruction. The other category is retail destruction. Let's consider each in turn.

Mother of Destruction. Wholesale destruction requires taking down the supporting societal infrastructure so that pretty much no systems or networks are going to function. Planning wholesale destruction depends a lot on the society that you are going after. Developed societies, for example, are resilient, hard targets. Take the example of the Great East Japanese Earthquake.

At 2:46 p.m. on March 11, 2011, an earthquake occurred 80 miles off the coast of Honshu (Japan's most populous island), approximately 240 miles from Tokyo. The initial shock measured at a magnitude of 9.0 on the Richter scale (making it the fourth most intense quake in recorded history). The quake was followed by powerful aftershocks, the first of which occurred only thirty minutes later at a magnitude of 7.4. Following the quake, a massive tsunami swept across the northeast coast of Japan, reaching several miles inland and flooding hundreds of square miles of land (including forty-two municipalities in four prefectures). With destruction and damage to roads, bridges, ports, railroads, buildings, and other infrastructure, as well as more than twenty-eight thousand people dead or missing, the full disaster caused by the earthquakes and tsunami affected more than two dozen prefectures with a population estimated at over fifteen million. The Miyagi, Fukushima, Iwate, Yamagata, Ibaraki, Chiba, Akita, and Aomori Prefectures were affected the most, and estimates of the cost of destruction are between $122 billion and $305 billion (between 2.2 percent and 4 percent of Japan's gross domestic product). In addition to this destruction and loss of life, facilities at the Fukushima Daiichi nuclear power station were severely damaged in the disaster, creating a national panic over the risk of exposure to radiation.[4]

For all the catastrophe heaped on Japan, destruction on par with what countries experience in war, Japan recovered from the disaster extremely well. That is because developed societies have more robust and resilient infrastructure and resources. They are just a lot harder to take down.

As a result, if a competitor wants to destroy networks on a large scale against powerful competitors, they must think about massive destruction. Short of murdering everyone or blowing up everything, there are only two means of wholesale destruction that have dramatic consequences for networks. One is an electromagnetic pulse (EMP) strike. The other is a cyber–Pearl Harbor. Let's look at each of these nightmare scenarios in turn.

EMP. Developed societies can pretty much weather anything except the widespread and sustained loss of electrical power. Without electricity, virtually no major infrastructure can function for long or at all. An EMP attack is one of the few ways that this level of destruction can be achieved.[5]

In the event of a massive EMP strike, communications would collapse, transportation would halt, and electrical power would simply be nonexistent. If it happened in America, not even a global humanitarian effort would be enough to keep hundreds of millions of Americans from death by starvation, exposure, or lack of medicine. Nor would the catastrophe stop at US borders. Most of Canada would be devastated, too, as its infrastructure is integrated with the US power grid. Without the American economic engine, the world economy would quickly collapse. Much of the world's intellectual brainpower (half of it is in the United States) would be lost as well. Earth would most likely recede into the "new" Dark Ages.

An EMP can be produced by the detonation of a nuclear weapon at high altitude or as the result of unusually powerful solar activity (often called severe space weather). An EMP generates a high-intensity burst of electromagnetic energy caused by the rapid acceleration of charged particles. A wave of EMP creates three chaotic effects. First, the electromagnetic shock can disrupt electrical devices. The second effect is like lightning—a power surge that would burn circuits and immobilize electronic components and systems. The third is a pulse effect that flows through electricity transmission lines, damaging distribution centers and fusing power lines. Any of these can cause irreversible damage to an electric grid system.

I described this threat in detail in *Wiki at War*. For years, I even promoted the idea that America needed a National EMP Awareness Day. The US Congress has long deliberated on this threat but not really addressed it. Federal agencies remain mostly ambivalent. So my idea was every year in Congress and the government for one day, everyone would shut off all their lights and digital devices to feel what the impotence of being on the receiving end of an EMP strike would be like.

Yeah, nobody cared.

Well, almost nobody. *Wired* magazine covered the nonevent in an article titled "E-Bomb Awareness Day: Grab Your Tin Foil Hat." The reporter, Nathan Hodge, wrote "That's right, Heritage is proposing a special day to raise awareness about the threat from electromagnetic pulse attack. Electromagnetic pulse weapons—what our own Sharon Weinberger dubbed 'the boogeyman bomb'—are the favorite doomsday scenario for national-security scaremongers."[6] Wow, that hurt. Still, it doesn't change the reality that an EMP is one of the few ways to ensure the mass destruction of networks in a modern society.

Cyber–Pearl Harbor. Is a cyber–Pearl Harbor even possible? Not sure. But if we had a cyber–Pearl Harbor, the results would be very bad. If there is no electricity, for example, there is no internet. Of course, that would only be one of many insurmountable problems we would have to solve. Beyond the certainty that if an attack could be pulled off it would be devastating, there is a lot of controversy over whether a massive cyberattack is possible. Could anyone deliver an attack that would result in long-lasting and widespread outrages? Who knows?[7] Absent an end-of-the-world disaster (like a planet-killing asteroid), no one can reliably place odds on whether anything like a complete shutdown of the internet by human design is ever going to happen. Here is what we do know.

We know what the internet is. It is not a thing at all. It is not owned by anybody. It isn't even really an "it." Basically, the internet represents the global capacity to transmit digital information through a commonly used system. The Internet Corporation for Assigned Names and Numbers (ICANN), a non-profit private company, supervises core functions, including assigning and managing domain names and internet protocol (IP) addresses. The Internet Assigned Numbers Authority (IANA), a department of ICANN, manages the Domain Name System root zone, the internet's "address book," which consists of thirteen sets of root servers comprising hundreds of servers at over 130 locations dispersed among many countries around the world. The internet's address book and a standard means to identify little packets of information and ship them in little digital envelopes is pretty much what makes the whole thing work. The vast network manages all the stuff that goes on behind the curtain so that when you google "meatloaf," up pops a recipe or the singer.

The infrastructure that stores and carries this information boggles the mind, spanning many countries with a lot of technical equipment involved.

Although decentralized, cyberspace remains dependent on the physical network of computer servers, fiberoptic cables, satellites, and an immense system of cables that have been laid across the world's oceans.

In addition to the internet, there are many other computer networks that we depend on every day. These are called supervisory control and data acquisition systems. They run or monitor everything from mass transit systems to electrical grids to all the processes on an assembly plant factory floor. Many of these systems are also hooked up to the internet.

So we know the internet and all these systems are big and important. We also know in mass they are resilient. There have been attacks on the internet by viruses (malicious computer software) that have gone global. There have been widespread outages. There have been attacks on the internet of whole countries, including nations like Estonia and Georgia. There have been unprecedented peak usage demands on the internet while parts of the internet have been disrupted. Yet, the internet still works. The lights still go on. The trains run on time (most of the time).

We also know that we are all dependent on the internet. If any world power managed to take down the entire internet, the resulting global disruption might hurt those malefactors as much as anyone. That also happens to be a bit of a deterrent.

On the other hand, there are some states that are cyber superpowers, like China, Russia, and Iran, that don't like the United States and its allies and can do some terrible things to us without damaging their capabilities too much. The United States also has "covert" cyber-capabilities that we are told are fearsome. Then there are small cyber powers that can also sting hard when they want to. There, for instance, is the case of North Korea hacking Sony Pictures (2014) and forcing the company to cancel the screenings of a film lampooning the "Great Leader." The fact is, there is a lot of dangerous cyber powers out there.

Who is to say someone won't try to take down a whole society's infrastructure with a massive cyberattack? What can be said for sure is that if they pull it off, it would be one of the few means to confidently ensure the large-scale destruction of networks.

Retail Destruction. Instead of taking whole societies offline or out of action, competitors can target individual networks for destruction. This is the tactic I like to call retail destruction.

How and how much it takes to destroy a network depends on the network. Take the example of al-Qaeda. After 9/11, the United States intended to not just defeat but destroy the transnational terrorist network. The United

States launched a major military campaign (2001) that succeeded in defeating the Taliban and al-Qaeda, driving them out of Afghanistan, but al-Qaeda was not destroyed. The United States then focused on eliminating al-Qaeda leaders, believing if the core leadership were taken out, the network would collapse. Yet, after capturing and killing thousands of high-profile al-Qaeda figures, the network didn't thrive, but it survived. After the Taliban returned to Afghanistan (2021), al-Qaeda leadership were spotted hanging around Kabul with Zabihullah Mujahid and his pals. Despite the determined American effort to eliminate them, al-Qaeda has proven extremely resilient.

The more resilient the network, the more difficult the destruction option becomes. That raises a second issue: Does the network's enemies have the resources, time, and the will to secure its elimination? Answering this question goes back to a topic that has been addressed before—the asymmetry of interests and resources. Destruction requires the attacker be willing to outmatch its enemies in both. For instance, during both the Korean and Vietnam Wars, there were proposals to use nuclear weapons to destroy enemy formations. Both times, the United States proved unwilling to take that step even though it had the means to do so. Washington lacked the will to risk such an extreme measure, even when the alternative meant the United States might well lose.[8] In contrast, during Operation Pointblank, the allies proved willing to flatten whole cities to knock out the German ball-bearing factories—and still, they failed. Against a resilient, determined competitor, destruction is a very serious business. The choices may be anything but easy to make, impacted by a range of issues from morality and geopolitics to means and practicality.

How to destroy a network can also be a difficult question to answer. It may not require killing everything in the network, but it does require eliminating the network's capability to function. That sounds easier. Still, it might not be easy. Belarus, for instance, has some of the lowest access to internet freedom in the world. During and after the controversial 2020 election, the government restricted access and jailed journalists and activists all with one goal—destroy the groups opposing the regime. However, denying conventional access to the internet has proven far from adequate in eliminating the political opposition harassing the regime. There are persistent reports of Belarusian activists finding unconventional methods for sustaining an online social media campaign against the regime.[9] The opposition, for instance, ferrets out embarrassing information, like the identity of KGB agents pretending to be pro-regime political activists, distributes the information on Telegram Messenger (which provides end-to-end encryption), and then figures out how to get the story into the public space. I know this happens because the people who help do it told me so. A strategy of completely denying political opponents total access to the internet, even in nations with restricted and monitored access, is extraordinarily difficult.

All that said, however, the reality is, if humans can build something, the odds are, other humans can find a way to destroy it if they are determined, resourceful, and powerful enough. Destruction of a network may be accomplished through different means, both physical and psychological. In the United States, for example, white supremacist groups have been eliminated by being financially bankrupted through court judgments.[10] Where there is a will, there may well be a way. As with other aspects of network competition, whether online or offline, don't just think of social networking warfare as a battle of "our electrons" against "their electrons." Scrutinize all the elements that allow a competitor to compete and how crucial they are to the functions of the network.

Offense and Defense

All the tactics discussed in disrupting and defeating adversaries are applicable for destroying them. The difference is in the measure of physical or psychological force that you are willing and required to apply to achieve the level of damage necessary to level a network, as well as the collateral damage (destruction on other than the directed target) that you might need to accept. It all comes down to choices, having the means to deliver devasting action and living with the consequences.

While fighting to inflict destruction may be a significant undertaking for the attacker, fending off catastrophe can be an equally daunting task for defenders. Infrastructure and networks, however, can be safeguarded against cataclysmic loss. Doomsday is not a foregone conclusion. Just as networks can defended against defeat (see previous chapter), they can be protected against destruction.

How much defense against devastation is enough? As with other aspects of competition, risk assessments are a key determinant in making investments to forestall destruction. How much is enough? The level of effort and resources that should be invested in deterring or dissuading catastrophic attack, protecting infrastructure or individual networks, or mitigating against destructive effects is determined by the value of what is being protected. Let's explain that by looking at an example from military competition.

Strategic defenses (missile defenses) are often dismissed as a practical means for defending against destruction by a nuclear attack because it is cheaper to build and field offensive nuclear weapons than defenses against them. That, however, is a flawed way to assess the cost, benefits, and risks associated with nuclear warfare. The value of missile defenses is not measured against the cost of threat systems—it is measured against the value of what the missile defenses are protecting. In this case, the value of the US economy is about $23 trillion. Furthermore, about 330 million people live in the United States. That is something that justifies a lot of missile defense capability.

When the functions of networks impinge on vital interests of national security, the prospects that destruction may be seen as a realistic option for one side or another may come more into play. In these instances, competitors will have to consider the role of offensive and defensive operations more seriously in networking warfare. What serious competitors who are facing serious competitors should never, ever do is assume the enemy will never do the unthinkable—just because our side doesn't want to think about it. When the stakes are high, be prepared to be a high-stakes player.

Weapons of Mass Destruction

Worry about weapons of mass destruction (WMDs). They are called that for a reason. Whether they are conventional, chemical, biological, or nuclear weapons, they are among the handful of destructive instruments capable of inflicting truly catastrophic damage on an unmanageable scale. As networking warfare becomes more central to matters of national security competition, there is every possibility that competitors will envision a role for them to secure their vital interests. Might there come a day when nations threaten nuclear war over cyberattacks? Who knows?

WMDs may be a factor of competition even if they are never used. The threat of WMD attack may be enough to blackmail, cajole, deter, or cower an enemy.

One key factor in this aspect of competition will be how widely these weapons are proliferated in the future. Who has access to them and what kinds of countermeasures are available? The course of counterproliferation (blocking or reversing the proliferation of WMDs) or nonproliferation (ensuring WMD technologies are not available) will significantly impact how this competition unfolds.

The proliferation of these technologies is, in fact, primarily another networking problem. Countries that have these capabilities are anxious that others don't get them. "Stealing or buying a ready-made weapon is a next to impossible feat," writes Togzhan Kassenova, a fellow in the Nuclear Policy

Program at the Carnegie Endowment. "The main path to a WMD is to procure components, material, and technology and then build a weapon. Because most of the goods employed in a WMD program are dual-use in nature, with indispensable civilian purposes, they are available on the international commercial market."[11] While nations impose sanctions and export controls to try to prevent unsavory actors from obtaining enabling WMD technologies, state and nonstate actors continue to attempt to use commercial and financial networks to secure materials that will allow them to build the weapons with which they can threaten total destruction. How this jockeying plays out will determine how widely these weapons will be available and how they might be used.

Weapons of Mass Disruption

There is also good reason to pay more attention to strategies and actions to immobilize and destabilize states, nations, populations, and large groups through large-scale disruptive social change. This category of destructive action could take several forums, and we are seeing plenty of examples in our contemporary world.

Mass Migration. There is a documented history of adversaries deliberately pressing forced migration as a policy of unrestricted warfare.[12] Mass forced migration can be implemented in two ways: (1) a "straightforward" threat to overwhelm a nation's physical or political capacity to prevent illegal migration; (2) through "norm-enhanced political blackmail" that restricts a nation's capacity to close its borders or restrict migration.[13]

These weapons are being used today. For instance, Russia has attempted to use illegal immigration to destabilize western European nations. In recent years, several Baltic and central European nations have accused Russia of intentionally attempting to flood their countries with illegal migrants from the Middle East.[14]

This weapon is particularly worrisome development with dramatic shifts in demographics with explosive growth in parts of the world that lack long-term stability yet because of geography and infrastructure are vulnerable to being exploited for forced migration. Africa, for instance, is currently growing its population three times faster than the rest of the world, widely projected to double by 2050 to about 2.5 billion.[15]

Mass Political Mobilization. From terrorist attacks to inspire or trigger catastrophic political disruption to mass political demonstrations, social and political action is another force in the contemporary world being harnessed to drive destructive change. The threat of and response to transnational terrorism after 9/11 is well told and well known. Recently, however, we have also seen powerful examples of mass political action.

In the wake of the October 7, 2023, attacks by Hamas and the Israeli military response, for instance, a global campaign sought to delegitimize the Israeli nation. This effort, which resulted in protests worldwide, was the latest attempt to use the instrument of mass mobilization to undermine the Israeli state. As one 2020 study concluded, in "the last two decades, international delegitimization of Israel has become a new mode of operation for those denying Israel's right to exist. It encompasses a wide range of civil-society and grassroots organizations."[16] In contrast, Israel has been accused of attempting to undermine the legitimacy of Palestinian nationhood.[17]

Genocide. Physical mass killing, sterilization, imprisonment, enslavement, and cultural annihilation remain a brutal part of the contemporary world. Perhaps most notable is the effort of the Beijing government to exterminate the Uyghur minority in China.[18] The ongoing conflict is another case where mass violence against a population has become a norm.[19] The willingness of modern powers to win by killing or disbursing populations cannot be discounted as real threats in the world in which we live.

The World Where We Fight

The implications of this chapter are sobering. That's why I started by invoking Pete Mansoor, the kind of thinker who always gives these kinds of issues the seriousness they deserve. He sets a standard for sober reflection on the consequences of conflict that networkers who are truly serious about networking warfare will have to match.

This chapter completes the section of *Digital Dominance* that explores how to go after an opponent's network. Destruction is a means for eliminating competition when other options are considered inadequate and costs and risks of delivering a catastrophic blow to an adversary is justified. There are two ways to eliminate a network. One way is macro-destruction that cripples the underlying infrastructure that allows society to function. The other way is retail destruction that makes a particular network inoperable.

Together with the previous section that detailed how to build a dominating network for national security competition, *Digital Dominance* has laid out all the components of fighting social networking warfare. The chapters that follow look at the landscape of competitors that are out there and what they have to fight with. *Wiki at War* surveyed this terrain over a decade ago, but much has changed since then. So here comes an update.

PART FOUR

World of Networks

11

TO WIN SUCH A WAR

Erin Walsh is a warrior.

I first met Erin when we worked together on the 2016 presidential transition team (discussed in a previous chapter). At the time, before the election, there was a cadre of exactly three of us working on the State Department: Erin, Catharine O'Neill (a terrific volunteer assistant who went on to work in the State Department and performed brilliantly), and I. I was lucky because Erin was my other half. As we crafted our advice for the new administration, I realized that when it comes to geopolitics, few people understand the United States' friends and enemies better than Erin.

Elizabeth "Erin" Walsh has a résumé that most Washingtonians would envy. Erin may not have ever been on a battlefield (though she did work in Bosnia for eighteen months before the 1992–95 war in Bosnia), but arguably, nobody has more experience in global competition. She served in four presidential administrations. The list of her jobs includes the White House Office of Political Affairs, the Department of State Near Eastern Affairs Bureau, the US Mission to the United Nations, the Department of Energy Assistant Secretary of Commerce for Global Markets and Director General of the US and Foreign Commercial Service, and Associate Director for Presidential Personnel as well as in the National Security Council as deputy assistant to the president and senior director for International Organizations and later for Africa Policy.

In addition to her government service, Erin spent a dozen years in the private sector working on global investments and in the technology space. She has seen the world from all sides now—government, civil society, and business perspectives—working here and abroad.

More recently, Erin Walsh joined our think tank where she committed herself to one of her greatest passions, a passion that emanated from decades of watching the post–Cold War world evolve: the challenge of China. China, Erin concluded, wants "to create their own new rules, and new economy, and new standards, and new world, for that matter; for the world to play by the rules that they set forward."[1] She believes this is bad for the interests of the United States as well as its friends and allies. Erin committed herself to doing what she can to stop them. And, she decided, a network was the best way to do that.

Erin Walsh has a flair for understanding both networking and strategy, combining the two in her efforts to counter China. Her idea is to focus on China's greatest weakness—its global reputation. Beijing made great advances in its quest for world power by selling the world on the idea that what Beijing wanted was good for the world. The more China does, however, the clearer it is that that is just one big lie, and Erin is out to expose that.

One project that she pioneered was an independent national commission to investigate Beijing's role in the origins of the COVID-19 pandemic. Her work was a textbook case of networking, successful because she recruited the right mix of talent—bipartisan trusted political and policy leaders, as well as credible scientific, medical, and forensic experts—and launched them on purposeful mission.

Erin's commission also benefited from her immense expertise in geopolitics. She is anything but a novice when it comes to understanding the Chinese Communist Party (CCP) and how it does business (more on that to come). She is, in practice, the poster child for the modern networking warrior who practices Sun Tzu's immortal adage, "Those who understand their enemies and themselves will always win."

Spotlight

The great weakness of the CCP and their oppressive and destabilizing regime is transparency. The more the world understands what the Chinese regime is up to, the more the world is not going to like what it sees and the more likely the world will do something about it. The prospects that this could happen are very real. There are think tanks and researchers all over the world conducting what is called open-source research—in other words, collecting publicly available information to reveal exactly what the regime is doing at home and around the world.

We already mentioned Erin's work on the COVID-19 pandemic. Take another example, the Uyghur genocide (mentioned in the last chapter), a concerted government effort to persecute the Turkic Muslim ethnic community in Xinjiang, China. The regime has incarcerated over a million, enforced birth control, abortion, and sterilization, imposed slave labor, and committed other atrocities. Much of the global outcry against the Chinese actions started with research conducted by think tanks and investigations by other nongovernmental organizations (NGOs). For instance, the Australian Strategic Policy Institute developed a database identifying almost four hundred facilities associated with the detention and persecution of Uyghurs.[2] In 2021, the US government made a public determination validating these abuses.[3]

The Uyghur genocide is not even close to being the only threat the regime poses to its own citizens and the rest of the world. Many in the world are still not fully aware of what China does. That was why we decided we needed to lead a transparency campaign.

Although all the work of exposing China's activity is laudable, it is not globally accessible. Our think tank thought the impact of all the ground-breaking research being done would greatly be amplified if it all could be collated to provide a comprehensive view of Chinese behavior and to make this body of work more widely available to analysts and policymakers around the world. Our idea was to create the China Transparency project, bringing together the folks doing cutting-edge research on the regime and sharing their work.

Within a year, we had established an active network of over a hundred world-class teams doing open-source research on China. From that collaboration, our team developed an annual report to evaluate the transparency of Chinese activities in eight key areas from their record on domestic human rights to foreign direct investments and military activities worldwide. In addition, we created a podcast called *China Uncovered*, which broadcasts interviews of researchers on their work.

What is most remarkable was that we started at the height of the COVID-19 outbreak when virtually everything was virtual. We also started with virtually no budget and no staff. What we did have was the power of networking. The team built one of the most comprehensive coalitions in the world studying the actions of the Chinese elite and making their activities transparent to the rest of the world to see for themselves. In fact, the results of this work was what convinced us we had to go out and hire networking warriors like Erin to do even more to take the fight to Beijing.

Competition

China is only one of the nations that is a major competitor in the networking space. This chapter will review them, updating what was reported in *Wiki at War*. The approach here, however, will be a little different. All the activity observed over the last decade, by attackers and defenders, is adding up to something—a clear pattern of how nations compete and the role net-working warfare plays in their designs. Rather than focusing on what these competitors are doing in the networking world, this chapter steps back and reveals why they are doing it and how they are trying to operationalize their interests. This assessment was informed by the framework for explaining social networking competition developed for *Digital Dominance*. With the playbook for how to build dominating action-oriented networks for national security in hand, it is much easier to understand what major players are doing—why some are on top, why some are losing, and why some are tread-ing networking water.

Troublemakers of the First Order

When looking at global competition, it makes sense to start with what is called great power competition (first mentioned in chapter 9)—the actions of China, Russia, and Iran. Great power competition is the new term of art that has become commonly used to describe the age we are in (the National Security Strategy of 2017 featured the term), as the phrase "the Cold War" was used to define the era of global competition between the United States and the Soviet Union.

Granted, the term "great power competition" is not precise. All the powers in great power competition are not great powers. Furthermore, not everyone agrees with the descriptor.[4] Still, it is just a label. Most historical labels are imperfect. After all, the Thirty Years War lasted more than thirty years. The Cold War was not a fight against hypothermia. Great power competition is just another historical bumper sticker so that we all know what we are talk-ing about.[5]

The reason China, Russia, and Iran are so often the focus of attention is geopolitics. Outside the United States/Canada, there are three important parts of the world that link the world together, three big links in the chain of globalization. These parts of the world are the Indo-Pacific (Asia), the Greater Middle East, and Europe. If these regions are generally prosperous and peaceful and the "commons" (lanes of air, sea, and cyber connectivity) that connect them are generally free and open, the world remains connected. Lose a link to war, conflict, or chaos, and globalization breaks down. China, Russia, and Iran are countries of concern because they are powerful enough and their actions are serious enough to threaten the stability of these parts of the world. That's why they are particularly worthy of the world's attention.

China, Russia, and Iran all have some common features that are worth noting. For starters, they each have dominant influence over the major ele-ments of their national power: the economy (including infrastructure and private enterprise), political structures, and the military. Less authoritarian societies do not maintain centralized controls over all aspects of national activities, particularly in the private sector and civil society.

China even has an explicit doctrine governing the management of national power called "civil-military" fusion. In China, the CCP considers every instrument and asset of the nation responsible to the authority of the CCP. The Chinese military, for example, reports directly to the leaders of the party, not the government. Indeed, it is not only the government instruments that come under their authority; everything does. As China scholar Dean Cheng notes, a "critical difference with the West is that China has never developed a separate 'civil society' outside the reach of political power."[6] This is important because it means that potentially when China fights with networks, it can fight with everything China has—everything, every weapon from the cash in its coffers to millions of individual Chinese hacktivists and a fleet of aircraft carriers.[7]

While Russia and Iran do not have the same comprehensive legal and societal structures as the Chinese regime, in practice they are authoritarian powers with tremendous sway over the exercise of power in their countries—power that can be coordinated and turned on enemies at their will.

Another feature shared by these regimes is their enmity toward the West (loosely defined as any nation that shares the values of the free world). They not only see Western powers as obstacles to the expansion of their power but also as threats to their legitimacy and influence. These are civilizational differences.[8] Despite the many variations and disagreements, Western nations (which includes countries all over the world with like-minded belief systems) hold a common belief in human rights, freely elected governments, and free market enterprise. China, Russia, and Iran believe in none of those. In fact, they see them as obstacles to their power and authority. Their goal is to eliminate that threat by undercutting their competitors. They will work hard to achieve that end.

What they also have in common is how they want to win. What they all share is a strategy—to "win without fighting." That doesn't mean they are not willing to engage in some fighting; but overall, they seek to accomplish their goals by not fighting a major military confrontation that would be as devasting to them as it would be to their adversary. That these powers would prefer to win without a cataclysmic confrontation leads me to think the option of wholesale destruction (described in the last chapter) to take down networks is less likely. I am also confident making this prediction because if I am wrong, who will be left alive to read this book and complain? Nevertheless, even though they may not pursue end-of-the-world scenarios, these nations will seek to maximize alternative means to undermine their competitors, achieving advances through disruption and defeat, rather than expansive, violent, protracted conflict and destruction. This has significant implications for networking warfare as well since networks can be highly effective tools for going after an enemy without starting World War III.

This summary suggests that networking warfare will inevitably be a seminal feature of great power competition. That is unlikely to change anytime soon. There is plenty of evidence out there that this is exactly what is happening. Let's break it down with some specific examples.

Dragon Rising

With the world's largest population, economy, and an activist foreign policy, there is no question that among the great destabilizing powers China's actions are the most consequential. They are arguably the pacing item in global competition.

After the end of the Cold War, there was a protracted debate over "whither China?" The argument was divided between "panda huggers" and "panda haters." The huggers argued that as China modernized, it would integrate into the global system adopting international norms and in the long-term become a net contributor to global prosperity and stability. The haters believed the opposite, that as the CCP gained power, China would become more confrontational and assertive.

The debate began to turn in 2002 when an era of pragmatic economic reform in the country came to an end and the party increasingly exerted control over not just the economy but all aspects of society. "Under General Secretary Xi Jinping (2012 to present), there has been a continuing reduction in the economic space outside CCP control," notes Dean Cheng. "Moreover, while the CCP may have formally expanded the party's core concepts to account for the contributions of its most prominent leaders, including Xi Jinping, it remains firmly fixated on Leninism, especially the concept of the Leninist 'vanguard party.'. . . Whether in the economic, political, environmental, or religious spheres, there is no entity that is exempt from the scrutiny of the CCP."[9] Today, the aggressive and destabilizing policies of the CCP seem all too apparent. Now the panda huggers are hugging the panda haters. This turn of events raises an important question: How does China plan to use its unconstrained power to secure its interests?

Unlike the Great Middle Kingdom, which sought to secure its place by keeping the rest of the world out, modern China, particularly Xi Jinping, sees its security in dominating the international system. The principal tool for this quest will be through information dominance.[10] This has huge implications for networking war. What dominating networks do best is dominate information. So, no surprise China is in the networking business big time.

Previous chapters mentioned efforts by the Chinese regime to command domestic networks, but China's efforts are global. Let's take a deeper dive into one case in point—the effort of the regime to influence and exploit the US higher education system through a variety of networks.

China's desire to tap the well of America's ivory tower is no secret.[11] Its campus-based Confucius Institutes (university and district school–based programs funded by China to teach Chinese language and culture) have received much attention of late, but that is just the ice cube on the tip of the iceberg. Several other Chinese programs also exert influence and exploit US colleges and universities. Here is a snapshot of some of them.

Thousand Talents Programs. Beijing's Foreign Thousand Talents Program aims to attract "high-end foreign scientists, engineers, and managers from foreign countries." Invitations and advertisements to participate come directly from Chinese research institutions that manage individual programs. But those institutions report to and are overseen by the government and the party, which provides financial compensation for participation.

Chinese funding of Chinese Students and Scholars Associations (CSSAs). The Chinese government sponsors and funds events for CSSAs on US university campuses. Among other things, CSSAs provide services to help Chinese students adjust to life and academic activities in foreign countries. Those services range from finding housing and roommates to organizing study groups and community activities. But the influence of CSSAs can be far from benign. "I came to the U.S. and thought, 'Wow, great, I'm in a free country. Now I hope that everything is cool and happy,'" one student recalled. "But I found out that the government extended their control to even Chinese students in America."

Chinese Gifts, Contracts, and Partnerships with US Universities and Funding of US Universities by Chinese Enterprises. US universities enter contracts with Chinese sources to foster collaborative partnerships. Part of these activities include contracts with Chinese companies. A Bloomberg analysis of data collected by the US Department of Education concluded that in six and a half years (through June 2020), 115 US colleges received almost $1 billion in gifts and contracts from Chinese sources. And that's just the amount that had been publicly disclosed.[12]

The China Scholarship Council. A nonprofit organization within the Chinese Ministry of Education, the council funds academic exchanges, Chinese scholars, professors, and other researchers. It also provides scholarships for Chinese students pursuing graduate and postgraduate degrees abroad. A study by the Georgetown University Center for Security and Emerging Technology estimates the government currently supports between twenty-six thousand and sixty-five thousand students in the United States.[13]

China's efforts are not limited to educational institutions or even the United States. For instance, in our report on China, Erin Walsh highlighted Beijing's effort to coop the Vatican into not just advancing control over Christians in China but empowering China's influence worldwide.[14]

All these organization are networks intended to expand Chinese control and influence. Even when they are recognized as a concern, as was the case of the Chinese Confucius Institutes (many of which have been discontinued by US universities), the Chinese government has adapted and sought other means to achieve the same ends. What the example of the regime's exploitation of the US higher education demonstrates is the complexity, scale, and proactive nature of the regime's networking way of war. China is a deadly serious networking player.

Bear at the Back Door

There are many contrasting descriptions of President Vladimir Putin's Russia. It has been depicted as a declining or middling power. There are times when these definitions seem more apropos than others. The fortunes of the Russian economy wax and wane. Domestic political unrest is persistent, though the regime appears to be firmly in control. Furthermore, Russia boasts a powerful conventional military and a massive nuclear arsenal (counting tactical nuclear weapons far bigger than the United States), as well as an expansive security and intelligence infrastructure. Finally, even though Moscow may not enjoy Beijing's iron control over the society, the regime has dominance, influence, or access to most sectors of the Russian economy and society including the media, the church, and most notoriously, a vast criminal enterprise with formidable cyber capabilities.

Regardless of the state of the Russian state as a great power, Putin maintains an aggressive activist foreign policy focused on establishing a hard sphere of Russian control over the surrounding nations, dominating their security, politics, and economy. This buffer is intended to ensure regime survival by maintaining an unbroken geographic belt of security around the Russian state. Putin is a geopolitical Michael Myers, on a relentless mission, obsessed with controlling a perimeter against external threats, seeking every opportunity and advantage to extend his influence.

Putin's designs are inevitably intertwined with issues of great power competition because extending his influence comes at the expense of undermining US interests and European stability. Thus, like Beijing, Moscow envisions competition with the West as a zero-sum game where securing its interests can only be achieved at the expense of impinging on others.

Moreover, like Beijing, Putin relies on instruments short of major war (and an occasional small war) to expand his influence. Putin is above all a tactician. As opposed to Xi Jinping's effort to press an inexorable advance against the international system, Putin is an opportunist, constantly attempting to shape conditions creating weaknesses in his adversaries that can exploited. The Russian strategy relies heavily on networks for this effort. Several examples have been cited previously. One of the most illustrative, however, is the collection of efforts employed to undermine the independence of three former Soviet states—Ukraine, Moldova, and Georgia. Putin's efforts against them are a case of comprehensive networking warfare worth considering in detail.[15] The 2022 Russian invasion of Ukraine was not a bolt of lightning from the sky. Rather, the invasion was the latest step in a deliberate strategy to undermine the stability of western Europe through intimidation, subterfuge, corruption, and force on Europe's periphery. In fact, Erin saw exactly the same kind of meddling influence when she was in the Balkans in the 1990s.

Ukraine, Georgia, and Moldova all sit on the frontier of Europe and the North Atlantic Treaty Organization (NATO), but they are members of neither club. That makes them both less risky targets for Moscow and perfect platforms for launching further intrusions against the West. They have something else in common: Parts of each country are occupied by Russia. Moscow engineered the violent takeover of Moldova's Transnistria in 1992, of Georgia's Abkhazia and South Ossetia in 2008, and of Ukraine's Crimea and Donbas in 2014. In each instance, the Russian occupation devolved into a frozen conflict prolonged less by ethnic divisions than by Moscow's determination to hold these territories hostage. Having established a foothold in each of these countries, Putin uses a variety of means including operations through covert networks, disinformation through social media, and illegal activity through criminal organizations to continue to disrupt and destabilize his neighbors. From armies of bots to stirring up trouble with Russian ethnic minorities in bordering nations, Putin is constantly looking for ways to unsettle his enemies.

Even before the latest war, Putin extensively used networks to counter efforts to stymie Russian influence. In response to overt hostility, each of these countries has turned to the West, looking for military partnerships and economic integration as protection against further Russian aggression and leverage in eventually regaining lost territories. Both NATO and the European Union (EU) responded with positive signals. Axiomatically, Putin then moved to dash any hopes of joining the West, heightening the sense of isolation and powerlessness. In every case, the Russians responded by making it clear that they would exert every effort to make European integration more difficult. Moreover, they vowed that the frozen conflicts would not be settled unless they are settled on Moscow's terms. Moscow then unleashed the weapons of networking warfare to ensure that further integration with the West becomes difficult, if not impossible.

What Putin did worked, likely prompting his recent effort to seize all of Ukraine by force. What had become apparent over the last decade is that the United States and Europe had grown frustrated with the lack of progress. Unable to see a practical way to break the logjam, they had mostly settled into a habit of treading water, alternately trying to get the Russians to be reasonable and constructively engage the three countries. With no visible effort to move the ball down the field, the people of these countries increasingly saw the drive toward Western integration as fruitless. Some argued it would be better to turn back to Moscow. To Putin, this looked like success and encouragement for more of the same. Even after his initial invasion of Ukraine was blunted, Putin shows no signs of doing anything different. So far, reversals on the battlefield have attritted his forces but not dampened his conviction. Likely, he will continue to harass his enemies with networking warfare as well as real war until he gets what he wants, to the detriment of the free and independent peoples opposing him.

Messing with Mullahs

Iran is without question just a regional power, but since the regime's actions have the capacity to destabilize both the Middle East and Europe and extend its influence to Latin America, it is a regional actor with the capacity to have dramatic global influence, directly threatening two of the most important parts of the world.

Iran has a lot to fight for and it fights hard. The objective of the Iranian regime is above all to perpetuate the promise of the Iranian Revolution (1978), which the elites of the regime largely define as the perpetuation of their own influence and power.

Although Iran lacks the power and reach of Russia and China, its geostrategic position makes the country an actor of consequence. Iran sits astride one of the principal routes of petroleum exports, in addition to itself being a significant energy exporter. Iran's borders also crosses and crisscrosses intractable tribal, religious, and territorial disputes that impact the stability of the whole region, not to mention that its nascent nuclear program is recognized as a grave threat for nuclear proliferation in the Middle East which could escalate with catastrophic consequences.

Iran is also a premier networking warrior. The regime has long been cited as the leading state sponsor of terrorism.[16] Indeed, the regime sees transnational terrorist activities and support for insurgencies, such as the war in Yemen, as an extension of Iranian foreign policy. Much of this activity is operationalized through networks from terrorist organizations like Hamas to criminal groups that smuggle weapons as well as other commodities. Iran cyber operations are a particularly noteworthy exemplar of how the regime seeks to extend its influence through networking warfare.[17]

The regime has been increasing its cyber networking activities in response to the vulnerabilities that have become apparent over the last decade. During the Green Revolution (2009), a series of protests that erupted across the country over disputed national elections, the government was forced to slow and block internet traffic to prevent social media from being used to coordinate and inflame demonstrations. Afterward, the government significantly increased its capacity to monitor and control internet activity. When outbursts occurred again in 2019, the Iranian regime once again initiated a nationwide internet shutdown. According to NetBlocks, which monitors internet governance by nations, the Iranian action was the "most severe disconnection" they had ever tracked "in any country in terms of its technical complexity and breadth."[18] The regime has invested significant resources in expanding its capacity to prevent social networking from threatening its control.

Likewise, Iran has been increasing its offensive cyber capability. "Iran and Israel are engaged in a not-always covert cyber conflict," points out cyber security expert Jim Lewis. "Stuxnet, a cyberattack on Iranian nuclear weapons facilities [2010], accelerated Iran's own cyber efforts."[19] Like transnational terrorism, Iran sees cyber conflict as extension of efforts to compete with adversaries including the United States. Iran already has an established track record of conducting cyberattacks against Americans, including using social networking methods to identify and exploit vulnerabilities.[20]

As long as the Iranian regime remains committed to the governance practices and objectives of the revolution it will likely remain in competition and conflict with the United States, Europe, and many nations in the Middle East. It will likewise continue to use the tools of networking warfare for conducting offensive operations against its adversaries.

Other Actors

Below the level of the great instigators are a host of nations that have significant networking warfare expertise and capabilities. These include, for instance, North Korea, which has robust cyber capabilities. The regime is one of the most aggressive cyber actors in the world, using its activities to bypass sanctions to generate income, conduct espionage, and punish its adversaries. "North Korea has conducted cyber guerrilla warfare to steal classified military secrets, absconded with billions of dollars in money and cybercurrency, held computer systems hostage, and inflicted extensive damage on computer networks," writes regional expert Bruce Klingner, who authored one of the most comprehensive unclassified assessments of the regime's actions. He notes that one senior US official even concluded "that North Korea was one of the top four cyber threats capable of launching 'disruptive or destructive cyberattacks' against the United States."[21] Social networking is an integral component of the regime's offensive arsenal. North Korea, for instance, frequently uses spear phishing and other malicious social media cyber practices to gain access to networks. I know for a fact this is true. I know because I have gotten some of their emails, all dressed up to look like perfectly legitimate messages from legitimate colleagues but really fakes designed to sneak malware into our information technology system.

North Korea is far from alone. Cyber tools and expertise are widely available. Harnessing them for networking warfare is not a significant challenge. The proliferation of technologies is common. FinFisher, for instance, is a spyware software developed by a private German company for law enforcement agencies. According to Citizen Lab, an interdisciplinary research group at the University of Toronto, at least thirty-two countries obtained the software to conduct domestic surveillance. Even very high-end capabilities are not out of reach on the open market for nations determined to field capabilities. In 2019, for example, the Department of Homeland Security detailed the case of the United Arab Emirates (UAE), which employed former US intelligence personnel working for DarkMatter, a cybersecurity company founded by Faisal al-Bannai, the founder of a UAE mobile phone vendor service. They were able to build for the UAE government a near-top-tier capability equal to those of the most advanced national security agencies.[22]

Advanced cyber capabilities are layered on top of the widely available social media platforms that are proliferating globally at an ever-accelerating rate. Only a few years ago, there were projections that access to the internet would triple by 2022 and that by 2030, almost 90 percent of the world would be online.[23] This turns the question of who to worry about on its head. The issue is not what adversarial threat might be capable of advanced social networking warfare online or offline. The question is, is there any that won't be a threat?

These states are still not in the league of the major players who can combine social networking warfare with other powerful elements of military, political, and economic power to present a truly significant threat. Nevertheless, although other nations might not be Conan the Destroyer, they represent real and present dangers that can't be ignored. This is certainly true for the United States. We know that because Americans are attacked online all the time. There were over one thousand major cyber breaches in the United States in 2020.[24] Not all of them were by states. Not all of them were the product of social networking warfare. But some of them were. Furthermore, these state actors are not just a threat to other states. In fact, they often target commercial enterprises and civil society.

I mentioned my think tank has been attacked by the North Koreans. The Chinese came after us as well; others too, I imagine. This is no secret. In 2020, the Department of Homeland Security and the Federal Bureau of Investigation issued a joint public warning. They reported "persistent continued cyber intrusions by advanced persistent threat (APT) actors targeting U.S. think tanks. This malicious activity is often, but not exclusively, directed at individuals and organizations that focus on international affairs or national security policy."[25] These actions come as no surprise. Adversarial states know nonstate actors can pry into their human rights and governance activities, such as our China Transparency project does. These efforts represent as much of a threat to them as other states. In contrast, commercial enterprises can also represent an opportunity, used as surrogates to evade sanctions or steal cash that can be used in the service of the regime.

There will be more on nonstate actors in the next chapter, but here, they are relevant to discuss in state-on-state competition because states bear some responsibility for their protection as well. In addition, some nonstate actors, like Spirit of America, contribute to US competitive efforts or support important humanitarian work.

Private sectors are also crucial to national economies. In the United States, for example, the internet sector accounts for over 10 percent of gross domestic product (GDP). Conversely, in 2020, losses from malicious activity online (from both state and nonstate actors) amounted to about 1–4 percent of GDP.[26] The fact is, online networking is big business. States must care about online and offline networking activities that affect everything from national productivity to tax revenue. Additionally, most of the critical infrastructure (like the electrical grid) in the United States, the backbone that allows the society to function, is not in the federal government's hands. This must also be safeguarded from adversaries. Thus, protecting domestic nonstate actors from threat states is also a crucial part of great power competition.

Furthermore, the networking offensive against state and nonstate actors is and will continue to be far more extensive than just the influence of adversarial states in the social media and cybersecurity worlds. Aggressors are active in many fields online and offline. One of the most aggressive and threatening forms of action is through commercial and financial networks, actions that the Center for International Private Enterprise labeled "corrosive capital."[27] Media, diplomacy, and other instruments of public influence are other prominent avenues for exerting pressure. This is often called "sharp power."[28] Another major effort is in influencing the policies and activities of international organizations. China is particularly active in this space.[29] Taken together, the assaults on liberal states are pervasive, significant, and ongoing.

The accumulated force of these threats represents a danger. Furthermore, there should be zero expectation that these state actors will tone down their aggression in state competition anytime soon—zero. It is just not going to happen. They will only stop if the regimes fall (and don't hold your breath waiting for that, though you never know, do you?). They will continue to do bad stuff and they will not back off. They will take advantage of those that don't protect themselves. The worst evildoers can do will only be thwarted if they really fear the retributions for the worst that they can do. Only states can deter states.

That said, these actors do not have a free lunch in networking warfare. States will fight back. More on that follows. But to be frank, probably adversarial states have more to fret from nonstate actors who will in the end likely be less constrained than the nations themselves in going after the malicious activities of adversarial regimes and their hateful actions against their own people and others (more on this intriguing aspect of competition in the next chapter).

The United States, Allies, and Partners

Social networking warfare is not strictly a standoff between white and black hats like the shootout at the OK Corral (but then again, neither was the gunfight at the OK Corral). That said, the states marked for assault by adversarial states are pretty much all in the same boat; they are the ones with a big target on their back. In their own self-interest, they must respond to defend themselves.

Great power competition has meaning for all states. Like adversarial states, the rest of the world is not interested in World War III either. Other states also want to win without fighting. Their conception of winning, however, is more modest. Broadly speaking, these states believe they can survive without the collapse of major aggressive powers. Rather, their goal is to safeguard against the destabilizing activities that would undermine their freedom, prosperity, and security. At best, they would seek to establish "open spheres" of competition that would allow for the relatively free and safe exchange of goods, peoples, services, and ideas, as well as the maintenance of the "commons" (the trade and communication routes) that connect the world together. At worst, they would be content to keep the cancer of authoritarianism from spreading.

Among these states, there is a diversity of opinions over how global stability in the face of great power competition can be achieved. Some argue for neutrality or isolationism (trying to avoid being caught up in the competition). Some argue for "regime change" (toppling enemies). Some argue for concessions and compromise (appeasing aggressive powers). Some argue that perhaps it's best to emulate what made adversarial states powerful (adopting authoritarian practices). The consensus, however, will probably not be any of these extremes. What is more likely is that nations will use the guiding principles of their national interest to identify what is vital to them and demonstrate sufficient will to secure those interests. A model of this approach, for example, was the common direction of like-minded strategic partners working for a "free and open Indo-Pacific," a strategy that even though it does not explicitly call for defeating China's regional designs for hegemony, it nevertheless commits to blocking Beijing from doing just that.[30]

The question is not whether states will conduct networking warfare in their own defense. They have and they will continue to. The question is how and how well will they fight? Let's look at that issue next.

America the Beautiful?

A good place to start is at the top. Although the United States is unquestionably the networking superpower in this bunch, it is also the one with the biggest challenges. The United States today is arguably the only nation with truly global interests and responsibilities. In addition, safeguarding the United States requires protecting a homeland of hundreds of millions of people and a multitrillion-dollar economy, not to mention working with allies and other partners to vouchsafe their interests.

Not only is the US mission daunting, but the United States has the unenviable privilege of being every adversarial state's adversary. This gives Americans fewer options than many might think. For instance, there is an argument that Washington cannot afford to give equal weight to every area of responsibility. My good friend Eldridge Colby, for one, argues in a very thoughtful book, *Strategy of Denial* (2021), that the United States must focus its effort on ensuring China cannot achieve hegemony over Asia and the Pacific. I agree—in part. The Indo-Pacific is important. Making priorities sounds reasonable. There is no scenario, however, where neglecting other vital interests makes sense. Neglecting vital interests is like a doctor telling a patient they have a bad heart, a brain embolism, and cancer and asking the patient, "Which one do you want cured?" If the United States loses the strategic competition in any of the key areas of global struggle, it loses. The United States does not have the luxury of focusing on just one adversarial state or one part of the world.

While the 2017 National Security Strategy took a robust approach to defending US vital interests against all comers, other administrations have taken a more restrained approach to protecting the United States. That alternative approach looks to achieve equilibrium with adversarial powers through managed opposition that looks to "compete where we must" but "cooperate where we can." That too is a notion that sounds reasonable in theory. There is little evidence to suggest, however, that this alternate approach can work in practice. Washington, Beijing, Tehran, and Moscow have no trouble finding places to compete, but there are, as it turns out, very limited and hardly any consequential areas where they proactively want to cooperate. As a result, no matter how timid or aggressive a US administration wishes to be, they have in the past and will almost assuredly in the future find no reasonable alternative to hard competing in an era of great power competition.

From a network warfare perspective, the limitations of the United States in great power competition have been discussed before. They are two.

The first problem (as discussed in chapter 1) is that the US government has the same problem every government has. Governments tend to be good at linear operations and less good at nonlinear activities. Networking Jedi tend to be good at both. People like Erin Walsh, creative, intuitive, risk-takers who can also be team players and team captains, are rare in government's top policy jobs.

That is not to say the United States doesn't play the field in the public and covert space. It does. The State Department, for instance, had a Global Engagement Center that fell under the Under Secretary for Public Diplomacy and Public Affairs. The center was tasked to "recognize, understand, expose, and counter foreign state and non-state propaganda and disinformation efforts aimed at undermining or influencing the policies, security, or stability of the United States, its allies, and partner nations" and to "inter-agency efforts to proactively address foreign adversaries' attempts to use disinformation and propaganda."[31] That is a big word salad that sounds impressive. The center produced useful analytical reports and direct some active operations mostly through federal grants.[32] Despite its oversized mission, however, in past years, the center didn't even use all the funds it was authorized.[33] A 2018 Inspector General report further concluded the center "did not have enough experienced personnel to issue, manage, and monitor cooperative agreements [grants]."[34] Even from a capacity perspective, the center was like a Hobbit challenging an army of angry orcs, unable to conduct sustained, impactful campaigns capable of changing adversarial behavior. It was also criticized for conducting partisan political activity. In 2025, the center was abolished.

Other federal agencies face similar challenges, as noted previously. The Joint Improvised Explosive Device Defeat Organization, for instance, had its fair share of challenges getting ahead of the insurgent improvised explosive device network in Iraq despite having at its height over three thousand employees and a budget of $4 billion. Apparently, just throwing money at a networking problem doesn't always produce results.

That said, it would be unfair to say that the United States' network warriors are the gang that couldn't shoot straight. Remember Kara Frederick? Okay, she did almost get her future husband killed in an aborted raid, but there were many other occasions where her team found their mark and dismantled more than a few terrorist cells. There was also the previously cited example of the Silk Road, a criminal network effectively taken down by federal and state authorities. There are many other cases where the United States has defeated all kinds of networks doing all kinds of nefarious things.

If there is a common thread in success, it seems that where efforts are decentralized and effectively share information across networks and are staffed with competent and skilled personnel, focused on their mission and not distracted by politics, they appear more adaptive and effective. Still, at best, these labors can win battles and maybe defeat a network or two, but they can't beat a determined state adversary.

The second problem is the structure of US power. Unlike authoritarian regimes, the United States can't easily fuse together all the elements of national power for conducting networking warfare. After World War II, the United States began a series of structural reforms starting with creating the National Security Council to better integrated federal efforts. Episodically, Congress and the executive branch undertake efforts to update these efforts, such as establishing the Department of Homeland Security (2002). Yet, it remains a challenge to be versatile enough to adapt and remain current with the social networking war.

An additional challenge for the US government is the problem of politics. Although it is understandable that policies and strategies of how to compete will change from administration to administration, it is also undeniable that changes are sometimes also imposed for political reasons which are distractions or detrimental to the efficiency and effectiveness of operational activities. One highly controversial practice in the United States, for instance, is the tendency to shift the focus of attention on domestic extremist groups based more on their alignment with political factions than on public safety risk assessments.[35]

Even if there was a unity and efficiency of government effort, federal action is limited in that the government does control most of the tools and players involved in the world of social networking warfare. Arguably, this limitation is also a strength in that unlike authoritarian societies, more decentralized liberal societies have a greater space for innovation, experimentation, risk-taking, and adaption, attributes that are well suited to operating in the networking world. That said, and it may well be true, it is also worth noting that authoritarian societies can also exploit these aspects of competition if they choose to. The Russian government often exploits the services of organized criminal networks that prove themselves entrepreneurial, innovative, and adaptive, including adopting some of the most cutting-edge cyber technology available.

The Arsenal of Democracies

The accumulation of states that ally (e.g., a formal treaty relationship like NATO) or partner with the United States (organize mutually beneficial policies, programs, operations, and activities) are numerous and global. Indeed, the distinction between cooperative allies and partners is becoming less relevant in the rough and tumble world of great power competition.[36] Relations are diverse and manifold. Rather than focus on alliance structures, let's zero in on three strategic regions where the good guys must cooperate to compete with adversarial powers.

Europe. There is NATO and the EU, and both have policies and programs related to network warfare. The EU, for example, funds the European Centre for Excellence in Countering Hybrid Threats. The center conducts research and programs that relate to networking competition. One report, for instance, examines ways in which China "presents systemic risks for liberal democracies."[37] NATO sponsors a strategic Communications Center of Excellence that evaluates various networking activities. One sponsored study assesses how China uses its social credit system to influence behavior.[38] These alliances, however, will not determine who wins a networking war. Neither NATO nor the EU have the institutional authorities or capacities to address defeating dominating players. The real power of networking warfare is in the national capabilities of the states. That's a challenge.

When it comes to great power competition against China, Russia, and Iran, Europe is all over the map like a blind Boy Scout without a compass or a map. Europe has no idea where it is headed. Some European powers describe great power competition like the parable of the elephants—when the elephants fight, the grass gets trampled. They don't want to be trampled. They don't want to get caught in the middle. They would rather be neutral in the competition with the United States and its rivals. Other states think that analogy is nonsense, holding that the Russians, Chinese, nor the Iranians respect neutrality and trying to be neutral means eventually one of them will mow you down at first opportunity. Furthermore, even the states that want to be neutral would admit they don't want to be a suburb of Beijing, Moscow, or Tehran either. Even then, however, they disagree on what nations to worry about. Southern Europe doesn't see the Russians as a threat. Much of central Europe does. Some worry about China. Some want to do business with China. Some want to do business with Iran, though no one really wants to have anything to do with the Iranian regime. Europe is a smorgasbord of geostrategic contradictions. What does seem consistent is that most states take networking warfare seriously and some are very good at it.

Estonia is a case worth looking at. As noted, the small Baltic nation fights above its weight in the cyber world, and cyber capabilities are important to networking competition online and offline. It might be expected that Estonia's massive reliance on e-governance and digitalization would make the country a tempting target for cyberattacks. It was. In 2007, the country was slammed with a widespread denial-of-service attack that struck government offices, the parliament, banks, businesses, and NGOs. Some called it "Web War I." In the wake of the incident, the Estonian government introduced a national effort to enhance cybersecurity. As one study of the Estonian effort notes, the country developed a deterrent posture against future large-scale attacks by developing layers of complementary capability. "A high degree of cybersecurity competence helps to defend domestic networks (deterrence by denial)," writes Piret Pernik, "and simultaneously contributes to Estonia's reputation as a leader in cybersecurity in the international arena, which helps to attain its cyber diplomacy goals (herewith, corresponding to deterrence by entanglement and norms)."[39] Pernik concludes that Estonia's efforts should be a model for other European countries.

Arguably, the foundation of a stronger coordinated response to networking warfare will come from bilateral (state to state) relations and other emergent formats for multinational cooperation (more than two states) rather than formal structural alliances like NATO and the EU. Such arrangements are more collaborative, adaptive, flexible, and better suited to the dynamic nature of social networking warfare.

Asia. The Indo-Pacific region doesn't have a formal alliance like NATO. This geographic space is also far larger, more populous, and more diverse. Rather than establishing a new NATO to deal with the likes of China and other geostrategic miscreants, what has evolved in Asia is a nest of strategic relationships that may be better suited for cooperation in networking warfare.

At the apex of these relationships is the Quad, an informal grouping of the United States, Australia, India, and Japan, four of the great like-minded Indo-Pacific powers concerned about Chinese behavior. My colleagues and I would claim a very modest amount of credit for the development of the Quad. This idea was first proposed by the George W. Bush administration, but it quickly went moribund. That was when the head of our Asia research team, Walter Lohman, suggested that we organize a Track II Quad.

Track I is government-to-government dialogue. Track II is NGO-to-NGO dialogue. We arranged our Track II effort with prominent think tanks in the three other countries with the common goal of pressing our governments to establish the Quad at first opportunity. We kept this effort going for years. When I served on the presidential transition team, I made a strong case for the Quad as well. This is not to suggest there is a Quad today just because of what we did, but the effort is noted here as another example of how networking can lead to action. What also provided impetus for the Quad was China and its aggressive behavior in the region and against these countries specifically from border and trade disputes to malicious cyber activity and "wolf warrior" diplomacy intended to intimidate other countries.[40] The fact is, the administration took the Quad proposal seriously, as did the other governments, and now the Quad is thriving.[41]

Beneath the Quad are a host of trilateral and bilateral efforts to foster cooperation among the countries. The US–South Korea relationship is a case in point. The association between the two countries, for decades, largely centered on joint concerns over North Korea. That has changed in recent years and continues to evolve. One area of expanded cooperation extends to Chinese cyber activity and other aspects of Beijing's networking efforts.[42] South Korea has a great deal of capacity and expertise to bring to the table as well as a real mutual concern for protecting its people from foreign malicious activity. This cooperation, in fact, prompted our think tank to undertake another Track II effort—a program with the Institute for National Security Studies in Seoul to work together on cyber issues.

The Greater Middle East. This region also lacks a formal treaty structure for mutual defense, intelligence sharing, and security cooperation. There is, however, the promise of a framework for security cooperation because of the Abraham Accords, an expanding series of agreements facilitating normalization of relations between Israel and the Arab states. This process could well lead to future mutual security agreements and economic integration that would better enable these states to cooperate with one another and the United States against a range of threats including network attacks.[43] I suspect even the vicious war started by Hamas in 2023 won't stop forever the process of collaboration between Israel and the Arab states. Time will tell.

Without question, one of the greatest assets in collective security will be the State of Israel. The Israelis, without question, are the cyber superpower of the region and in many respects world class in some of the capabilities in both the government and private sectors.[44]

Middle States

Painting groupings of liberal and authoritarian states requires a broad brush, but covering the states in the middle requires a paint roller. No matter how ambivalent their policies or governments, some of them can't be ignored. They are pivotal to great power competition. Indeed, states with even modest resources can be important because they have a key geostrategic position or can be net contributors in shifting the balance between great powers one way or the other.[45]

Most problematic for liberal states is crafting partnerships with mildly authoritarian states that are part of the great whitewash in the middle between brutal dictatorships and democracies. There is a long list of the states. I won't list them. I am hoping *Digital Dominance* is translated into a zillion languages, so I don't want to offend anybody (except the governments in Tehran, Beijing, and Moscow). I will discuss (mostly) in generalities.

Jessica L. P. Weeks tried to draft a guide for dealing with these middling powers in her book *Dictators at War and Peace* (2014).[46] She writes with good reason. Authoritarian regimes have always been a bit of a bugbear for the free world, particularly for the United States. On the one hand, America is the country described by Alexis de Tocqueville (1835) as the "exceptional nation," a nation founded on democratic practice. Americans find everything dictatorships stand for abhorrent. On the other hand, we are also the children of Niccolò Machiavelli (*The Prince*, 1532). The first duty of the government is to defend the vital interests of the state. The demands of national security, on occasion, require doing business with less savory regimes.

Figuring out how to deal with hard-to-handle nation-states has been the acme of US global leadership in the modern era. Our record is an admixture of success, failure, and questionable calls, from handling Imperial Japan before World War II to managing relations with apartheid South Africa. When history failed to end with the end of the Cold War, dealing with difficult non-democratic states became a chronic post–Cold War problem as well.

Every part of the free world the free world cares about has these states, so figuring out where they fit in the great power competition and networking warfare is important. Weeks suggests that we classify them and choose to deal with authoritarian regimes according to the way in which internal forces constrain the leaders who decide the issues of war and peace for them. She stresses the importance of determining whether there is a domestic audience that can hold the leader accountable, and whether the influencers and leaders are civilian or military.

Weeks's model doesn't exactly work because it assumes we understand the internal workings of these regimes. We often do not. In *The Case for Democracy: The Power of Freedom to Overcome Tyranny and Terror* (2006), Natan Sharansky not only does an excellent job of explaining the difference between free and unfree states but also highlighting why dictatorships are so troubling. Closed regimes are nontransparent. Accountability, responsibility, and authority are obfuscated by webs of corruption, disinformation, secrecy, oppression, and deceit. Thus, even though it may be true that not all evil empires are equally threats to their neighbors, making the call between the odious and the unacceptable can be harder than it looks.

Still, Weeks's basic recommendation is right. The United States or any liberal democracy should not expend a great deal of statecraft on states that don't matter. Take the example of the big bad powers, Russia and China. Nadia Schadlow and H. R. McMaster framed how to deal with Russia exactly right in the 2017 strategy. The US way to deal with Moscow shouldn't be about preparing for a new Cold War or trying to turn Russia into a democracy. Washington's primary goal is marginalizing Putin's capacity to conduct disruptive activities that threaten the peace and stability of western Europe. Likewise, as the 2017 strategy argued, the security relationship with Beijing can and should be structured to limit the potential for conflict without compromising US interests. That same careful crafting of statecraft ought to be applied to lesser bad boys as well.

But just because certain dictators might not represent a cause for war doesn't mean the United States should actively court them. War and peace are not the only issues at stake. Even if some brutal regimes (like Hafez al-Assad's Syria) pose a lesser threat to us, they are still evil. Authoritarian bullies consciously and intentionally suppress human rights and individual liberty. Making nice with them is dumb unless it serves a compelling national security interest and provides direct benefits to us. Expecting them to play nice is naive unless they have a compelling national interest to do so.

During World War II, few regimes were more despicable than Joseph Stalin's Russia. But we needed the Red Army to win the war, and Stalin needed us to survive. A deal with the devil, yes, but a deal worth making. In contrast, the Russian "reset," the kowtowing to China, the Iran deal, and lifting sanctions on Cuba were all predictably foreign policy dead ends. The United States got nothing substantive and in great abundance.

Likewise, the United States or any liberal democracy should be the last nation to say any dictator is a good thing for their compatriots or their neighbors. Assad is a case in point. The Syrian dictator is a mass murderer and a source of misery for many throughout the Middle East. But publicly demanding he step down is pointless because Moscow and Tehran support him; Assad isn't going anywhere. Furthermore, dumping Assad isn't crucial vital interest to anyone but Assad. So liberal states should keep him at arm's length and marginalized.

The bottom line is that for liberal democracies in their networking activities or other actions, there are times when promoting freedom makes sense, even when issues of war and peace are not on the line. For example, promoting economic freedom for unfree states is a sound idea, whether those states are adversarial to us or not. It's in their people's best interest and in ours.

So the challenge for liberal states is not just developing realistic ways and means for dealing with various petty dictators but to revitalize the narrative of freedom as well. Part of protecting the freedom, security, and prosperity of nations is respecting and promoting human rights and liberty for all, while defending the principles of state sovereignty. It is balance of effort that if done right can help add the ranks of the states fighting networking war on the side of the liberal world.

The Struggle Among States

Now that we have surveyed the field, it is time to ask the hard question: Are states any better agents in the social network world than they were a decade ago when *Wiki at War* measured what was going on? The answer is an unqualified yes. That's cold comfort. The state of the competition is much higher than it was a decade ago and the stakes are way, way higher as well. So arguably, even the fiercest competitors are no further ahead than they were when *Wiki at War* went on sale.

Here is where we are. States have demonstrated that they have the capacity to be better competitors, direct better coordinating efforts, build better bridging between linear and nonlinear operations—all stuff that really makes for the best networking warriors. Liberal states, in fact, have the potential to outcompete authoritarian regimes. Whether they will or not hinges on the decisions they make on a couple of key issues. There are four on the list.

Politics. As noted, the United States has struggled with the intervention of political agendas that overshadow national security priorities and inhibit practical, realistic action. The tension between domestic politics and national security is an inherent challenge of statecraft. Resolving these disputes in favor of the national interest is part of what makes nations better competitors. During World War II, for example, political advocates pressed on Franklin D. Roosevelt to make the New Deal (the president's plan for economic recovery from the Great Depression) an integral part of the war effort. FDR demurred. He needed Republican congressional support to mount a true national campaign to defeat Germany and Japan. Many Republicans were hostile to the administration's New Deal programs. If Republicans thought FDR was shaping the war effort for political advantage, they might not give the president the support he needed. FDR could not risk that. The United States and other liberal states will need equally disciplined efforts to keep their politics out of networking warfare.

Priorities. Unlike authoritarian regimes that have weak social contracts with the people they govern, liberal states offer the promise of ensuring the freedom, safety, and prosperity of their polity. Addressing all these equities is always challenging, particularly since they can often be at odds with another. As with other aspects of public policy, liberal states will have to find ways and means of conducting networking warfare where they can deliver more security that does not come at the expense of the freedom and prosperity of their citizens. The United States faced one exemplar of this challenge after 9/11 with the passage of the USA PATRIOT Act, intended to provide more authorities to combat terrorist networks, which critics complained established undue infringements of civil liberties. Equitably and efficaciously resolving the friction between the competing priorities of security, freedom, and prosperity is an essential task for good governance, a healthy civil society, and a resilient polity.

Partnership. In liberal states, many of the capabilities for networking are in the private sector. States have limited capacity to direct and regulate these assets. In many cases, giving governments more power might be a bad idea. What might make sense from a national security might not make good business for the companies. Furthermore, many of the most powerful private sector companies have an international footprint. What might be best for their business might not match national policies. Building effective partnerships is an important task for states. It is not always easy.

In the United States, for example, in 2018 Google stopped working for the Department of Defense on a major artificial intelligence program after four thousand employees wrote protesting the project. Amazon and Microsoft entered into a bitter contract dispute for providing cloud services to the Defense Department, which resulted in a multiyear delay preventing the department being able to purchase any support. More on the role of "big tech" in networking competition in subsequent chapters. At this point, it is just worth noting that it will be extremely difficult for liberal states to be world-class networking competitors without effective public-private cooperation.

People. *Digital Dominance* has emphasized repeatedly that winning is at its core a people problem. The human dimension of networking competition is the most vital part to get right. Recruiting, training, and retaining people with the right skills, knowledge, and attributes for national security networking is the most critical task for states to get right. The US government knows this well. For example, pretty much every federal agency struggles to sustain a sufficient workforce to conduct all their cyber activities.

The State We Are In

This chapter has highlighted how the emergence of great power competition has shaped developments in networking warfare, surveying the actions of key adversarial and liberal states (a way better assessment than *Wiki at War* in my opinion). The chapter concludes with some optimism that liberal states can effectively compete if they can address the key issues laid out above. That is the good news. The bad news is that most won't be able to adequately overcome the challenges that keep them from becoming world-class network warriors. That doesn't necessarily mean they will lose the great networking wars of the future. The slack in capability might well be taken up by nonstate actors riding to the rescue.

Can substate actors really step in and save states that can't save themselves? Maybe. The prospects for a happy ending to the cliffhanger confrontations of state competition are considered in the next chapter, which ponders the role the nefarious and the virtuous independent networks play in the networking world.

12

THE WORLD'S A STAGE

Lisa is a lion.

We met Lisa Curtis in the prologue. She was in the thick of the scramble helping get innocents out of Afghanistan. Even after the immediate crisis abated, Lisa remains an active and staunch advocate for the women of Afghanistan.

At the time of the crisis, Lisa served as a senior fellow and director of the Indo-Pacific Security Program at the Center for a New American Security, a position she took after serving four years on the National Security Council staff. That assignment capped over twenty years of government service on the Hill, in federal agencies, and the corridors of the Old Executive Office Building.

In 1990, Lisa graduated from Indiana University with a bachelor of arts in economics. That, by the way, is the same school Justin Rhee went to (see last chapter). Go, Hoosiers! Her sheepskin from Indiana is also the highest degree Lisa Curtis ever earned, a powerful reminder that what makes for great networking warriors is not a résumé but the right skills, knowledge, and attributes to network. It doesn't matter how you get them, just that you have them.

Lisa learned her profession on the job. She started as an analyst on South Asia at the Central Intelligence Agency (CIA) and later served at the embassies in Pakistan and India. Then Lisa worked as a senior adviser in the South Asia Bureau at the State Department. From there, she also served as a staffer on the Senate Foreign Relations Committee, working for Senator Richard Lugar.

After leaving the Hill in 2006, Lisa made her way to the think tank world.

Sure, Lisa is smart, talented, and at least it seems, knows every single living human in South Asia. Let me tell you, however, what makes Lisa special.

Lisa, her colleagues, and I traveled together a lot to South Asia. I observed Lisa in action firsthand. It did not matter where we went—India, Pakistan, Sri Lanka, or Bangladesh. Everybody respected and listened to Lisa Curtis. She didn't just open doors. Doors opened for her, everywhere we went, at every level of government. Here is what is most remarkable about that experience. For starters, let's be honest, in many of these places, women do not play a prominent role in foreign affairs and international relations. Lisa didn't break glass ceilings. She waltzed through them like they were not there. The other remarkable thing about Lisa is that she was welcome wherever she went; even as she traveled from country to country where the governments did not like each other, they all respected Lisa Curtis.

Lisa did many remarkable things while she was with us including helping organize our Track II dialogue of the Quad (see last chapter). This may, in part, explain the administration's enthusiasm for the initiative after she went to work in the White House.

It was not much of a surprise when H. R. McMaster offered her the South Asia policy slot on the National Security Council staff. Nor was it much of a surprise that she took it. It was also no surprise she was one of the longest serving staffers on the council, outlasting H. R. and two more National Security advisers that followed. She is just that good.

In her latest position at the Center for a New American Security, Lisa continues to do what Lisa does best—connect with people and get things done. Lisa was never the most prolific writer or public speaker. She is an extrovert compared to Nadia Schadlow (chapter 9), but honestly, that is not saying much. What Lisa does better than any networker I know is to connect with people, listen to them, reason with them, learn from them, and turn that into an action agenda. If there is ever a hall of fame for the champions of managing conversation-mode networks, Lisa should be the first one inducted.

If there is ever a bipartisan, sensible US policy to fix the mess America left in Afghanistan after our abrupt exit, I suspect it will be because of the behind-the-scenes networking of Lisa Curtis, working across the government and nongovernment spaces to turn responsible ideas into action.

If there is a better exemplar of the power nongovernmental networkers to change the world for the better, I don't know who they might be.

Evildoers and Saints

Nonstate actors, like Lisa Curtis, are significant global players in networking warfare. That's not new or news. The number of nongovernmental agencies (NGOs) exploded after the Cold War.[1] That ought to come as no surprise; absent the hard spheres of influence dividing the West and the Soviet bloc, there was just a lot more civil space to fill. The world has also become way richer and a lot more people have a lot more money to fund NGOs. Finally, technology has been a great enabler, none more powerful than social networking platforms that have created unprecedented opportunities for civic action from social activism to humanitarian operations. If all the world is indeed a stage, there are a lot of actors performing. Today, there are millions of NGOs. The estimates are over ten million. In the United States alone, there is an estimated 1.5 million.[2]

It is even hard for states to keep up with the rapid expansion of nonstate actors. Major adversarial regimes like China have a hard time managing. In the last chapter, *Digital Dominance* described our think tanks in the China Transparency project. One of the insights from our first workshop was that we learned that the biggest consumer of open-source information on China was—wait for it—the regime in China. Researchers described that after they published their research based on Chinese sources, the sources would be scrubbed from the internet. The Chinese government was tracking their critics and then covering their tracks. There is a game of cat and mouse between the regime and its investigators going on almost every day. What this dynamic demonstrates is that the regime must fight to keep up. Not even the most closed states on the planet are completely immune from the meddling of nonstate actors.

The Dark Ones

Wiki War has a lot about the many bad actors networking online, as does P. W. Swinger and Emerson T. Brooking's *LikeWar: The Weaponization of Social Media* (2018), not to mention a whole bunch of other books and studies. Here, we are going to add something new. What *Digital Dominance* is going to do is talk about is why and how they really matter to networking war.

There are three categories of malicious players online (some of them play in more than one league at a time). (1) Terrorists and other violent groups. This includes Zabihullah Mujahid and others with all kinds of blood on their hands. (2) Criminal networks, such as the Silk Road run by Ross Ulbricht, the "Dread Pirate Roberts." (3) Political extremists, like Richard Spencer and other organizers of the infamous 2017 "Unite the Right" rally in Charlottesville, Virginia (August 11–12, 2017).

Each of these types of groups may have different objectives and motivations. Some may cooperate or coordinate with state actors. There are two reasons why the ones that matter really matter.

First, these networks have the capacity to operationalize malicious action in the real world. The Taliban matter because they could overrun a country, brutalize women, and harbor no-kidding terrorists. Ulbricht matters because, by some estimates, he ran a criminal enterprise that did millions of dollars of illegal business.[3] Spencer matters because he and his fellow organizers sparked such a rampage that they got hammered by a jury in 2021 with a $25 million judgment for inciting violence.[4] In other words, why they matter (for our purposes) is not because they are evil but because they can generate troubling real-world action.

Second, they matter not necessarily because of the physical harm they can achieve (as awful as that might be). They matter because the result of their action can potentially undermine the confidence, credibility, or cohesion of liberal states. No matter how inept these states are at network war, the free world is not going to win, let alone survive, in networking war (or any other war for that matter) if they can't manage these marauders online and offline. In short, malicious networks that can harm a state's vital interests by harming the state's capacity to act are nonstate actors worth worrying about.

There is a reason why their capacity to undermine the credibility of the state is crucial. Credibility and cohesion are the coin of the realm that give national power its real value. Liberal states can't effectively manage these instruments absent the consent of the governed.

Here is an example from studies of disaster response that makes the point. Researchers at the University of Delaware's Disaster Response Center once conducted an interesting project. They interviewed people about how they thought people respond in disasters. Most answered that there would be panic and mayhem. What the researchers realized was that respondents were describing what they saw in disaster movies, like when Godzilla tramples Tokyo.[5] People with no real-life frame of reference fall back on popular culture to fill in the blanks. Then the researchers surveyed people that lived through a disaster. They concluded that how people respond is very different. People rarely panic or start acting like the little monsters in *Lord of the Flies* (1954). Rather, they act sensibly and rationally. That was true in most cases—except in cases where people sensed law and order had broken down or they could not count on the authorities to restore social order.

There are plenty of real-life illustrations of human response in action. One of the starkest is the contrasting response to two blackouts in New York City. Each lasted for a little more than a day. The great blackout of 1965 was considered a nuisance. In contrast, a similar blackout in 1977 saw a very different response. This blackout resulted in widespread looting and the breakdown of the rule of law throughout many New York neighborhoods. The estimated cost of the blackout was approximately $346 million, and nearly three thousand people were arrested during the twenty-six-hour period. The social order degenerated so quickly that *Time* magazine called it a "Night of Terror." New Yorkers acted differently in 1977 because New York was a very different place, plagued with high crime, poverty, extreme racial animosity, and social tensions. People had no confidence in their city. There were many explanations for the sudden violence in the aftermath of the blackout, with justifications ranging from racial hostility to cultural shifts, even to weather, but the simple fact is that during disaster, "'under stress' or 'exceptional circumstances,'" the poor saw "no reason to play by the rules." A single act of disruption triggered chaos.[6] Low social and political stability make societies, rich or poor, extremely fragile.

Nonstate actors with a significant capacity to undermine social and political cohesion and consensus or can exploit societal weakness in matters related national security and important vital interests are a real threat. This is, for example, why Vladimir Putin is constantly looking for ways to mobilize Russian ethnic minorities in neighboring countries, including through funding NGOs, and supporting formal and informal resistance networks. He knows these can be weaponized to good effect. One example is the Russian practice of surreptitiously funding "nonprofit" environmental groups in central Europe to rouse opposition to energy projects—that just happen to compete with Russian-sponsored energy projects. These kinds of assaults on a free society can have debilitating impact.

So which nonstate actors (for national security purposes) are the ones that are real causes of concern? The intelligence preparation of the battlefield framework is a handy tool for helping answer that question because one of the insights it provides is competitive assessment of what a malicious nonstate group might be able to achieve. Risk assessment can be another helpful analytical tool. What is key to the effectiveness of both, however, is (1) having sufficient information to do a realistic assessment and (2) keeping political and other biases out of the analysis (as discussed in the last chapter).

The reality is that in most cases, the results will likely show that resilient states won't find these actors in and of themselves as severe national security threats. That is because the states are resilient. Estonia offers a case in point. Before the unprecedented cyberattack in 2007, the nation was racked with two nights of riots and looting, after protests sparked by Russian-language media that resulted in the death of 1 person, 156 being injured, and 1,000 people arrested. That was followed by a massive hack attributed an ethnic Russian Estonian national who was convicted of launching the cyberattack as a protest against the government. What is notable is Estonia quickly recovered from the mayhem because of its political stability which manifests in a high degree of confidence in the national government.

In contrast, if a society is rocked by nongovernmental action, as was the case by the Iranian regime in 2009 and again in 2019, which requires a draconian government response to suppress public disorder, this is the result not of the influence of the NGO per se but of the fragility of the regime. In these cases, the state's real problem is not the power of the NGO but the weakness of governmental structure and the frailty of the nation's civil society.

None of this is to say that even the most resilient states should ignore malicious acts by nonstate actors. Malevolent acts, like the "Unite the Right" rally, are clear dangers to public safety. They should be addressed through the appropriate framework of public order.

Dark Ones with Guns

Heavily armed NGOs, capable of conducting military-style operations to advance their aims are a particular nonstate threat that merits special attention. These groups are often referred to as mercenaries or militias, though these terms are often used with great imprecision. Better to just call them armed actors.

For starters, let's make a distinction between what is and is not the problem. During and after the Iraq War (2003), much was made of the US use of contractors in combat, who were often referred to as mercenaries (under international treaties, mercenaries are paid soldiers who are not under the command of parties to a conflict. Contractors in Iraq are explicitly exempt; they were not mercenaries because they were employed by the US government). By 2007, there were over 100,000 individuals working on US contracts in Iraq and Afghanistan. By contrast, there were only about 160,000 US combat troops. Contractors were so ubiquitous, and as operations in both theaters became more protracted and difficult, how could their use not become contentious? Controversy was exacerbated by tragic incidents and scandals, including the use of contractors at the infamous Abu Ghraib prison run by the US military and the CIA where detainees were tortured and subjected to unspeakable conditions.

I became interested in the issue because a real debate had erupted over the use of contractors for national security missions. After studying the topic for a year, I concluded that the controversy was more about pushing political agendas than identifying a real problem to be solved. When contractors were used appropriately according to the laws, which were adequate to govern their activities, they were fine. Contractors were being used more in combat because the private sector had developed a much greater capacity to deliver services that were useful for governments fighting wars. Indeed, I concluded, rightly, that their use would grow, not shrink, in the future—and, if properly managed, were an asset, not a limitation.

Not wanting to waste all the research, I decided to write a book about it, *Private Sector, Public Wars: Contractors in Combat—Afghanistan, Iraq, and Future Conflicts* (2008). By the time the work came out, the controversy had largely moved on. A new president (who did not start the Iraq War) had been elected, and now the contractors were working for him. All of a sudden, most people didn't care about the issue anymore, even though they were the same contractors that had been working for the last president. The way the issue quickly collapsed as a cause célèbre seemed more confirmation that it had been played up for political advantage more than anything else. The book made the point that making better use of the private sector is potentially a powerful national security tool for the public sector. More on that topic in the next chapter.

The real problem of armed actors is nonstate or adversarial states hiring and organizing armed groups to conduct illegal activities. An assessment from the Brookings Institution (which has a major project studying armed actors) warns, armed "criminal and militant groups, as well as other nonstate armed actors, have become relatively stronger."[7] They are a particular threat in fragile states that lack the capacity to provide public safety and the rule of law.

More problematic are groups like Hezbollah in Lebanon and Hamas in Israel. Their organizations have both political and military wings. The legitimacy of their military activities is tied to the controversy over their political agenda. In many respects, they have to be treated like quasi-state actors. The difficult debate over the authority and response to these groups was illustrated by the global sparring over support for Israel and Hamas in the fighting following the October 7, 2023 attacks on Israeli citizens.

Adversarial states also fund armed actors to intimidate and threaten other nations. Russia, for instance, is often accused (and rightly so) of employing "little green men," surrogates used to destabilize Ukraine and other countries.[8] China directs a maritime militia to harass other countries in the South China Seas.[9]

The expectation that the world will see more of this makes sense, not necessarily because of the persistence of fragile and failing states but more because of the pernicious influence of great power competition. *Digital Dominance* made the case that great powers may not be determined to seek the destruction of their opponents; nevertheless, they will constantly seek ways to compete that undermine their opponents.

Great powers are, and likely will continue, taking a page out of Cold War competition. In 1957, Robert E. Osgood wrote *Limited War: The Challenge to American Strategy*. Osgood argued as the United States and Soviet Union stalemated making the prospects for nuclear and large-scale conventional war less likely that they would seek out other ways to undermine each other. Over the course of decades, proxy wars, insurgencies, and terrorism pockmarked competition in Africa, South America, the Middle East, and Asia. Today, we see similar activity in hot spots around the world.

While armed groups fight with bombs and bullets, like all organizations in the modern world, their activities and influence can be enabled or diminished by action-oriented networks. Criminal cartels in Mexico, for instance, control physical territory through armed force, expanding their influence and capacity through illicit activity including attracting customers for their human smuggling operations on Facebook.[10] A contrasting example is the global social media campaign in 2014 to recover kidnapped Nigerian schoolchildren taken by an armed group.[11] Inevitably, the activities of such groups will become intertwined with networking warfare.

Bright Shining Lights

Disruptive nonstate actors are not the only forces of consequence for states to consider. In the last chapter, *Digital Dominance* advanced the proposition that the decisive advantage in social networking will come from constructive nonstate actors who can operate faster, more creatively, and decisively than governments. Furthermore, unlike groups and organizations in authoritarian states, they can exploit the full advantages of free and open societies. States can extract the full advantage of what these private networks can accomplish by protecting free and open spaces online and offline that facilitate their activities.

Who are these guys?

Tanks that Think

During World War II, when military planners went to a secure location to draft plans to defeat Japan and Germany, they called these places "think tanks." After the war, the US government organized federally funded research and development centers to employ many of the researchers, scientists, and strategists they had used during the war to continue to address national security issues. The most famous of these is the RAND Corporation. Today, there almost two dozen of these private institutions that work for the US government. They inherited the term "think tank," but that nom de plume has also been applied to the constellation of private public policy research institutions that are neither affiliated with academic universities nor government research facilities.

The reality is that "think tanks" have been around long before RAND. They have, however, widely proliferated in the last few decades. For several years, the Think Tank and Civil Societies Program at the University of Pennsylvania produced an annual index and rating of these organizations surveying almost two thousand scholars, policymakers, donors, and grant-making foundations, ranking 6,500 think tanks in the world. The survey covers everything from expertise and impact (not bragging, but my institution was ranked the number one in the world for impact three years in a row), to the extent of their policy networks and governance. The most impressive takeaway from the index is that there is an arsenal of ideas and action out there that cumulatively rivals what states can do. There is a sophisticated and complex network of global think tanks that can impact the networking world online and offline.[12]

These organizations comprise everything from small offices of a few people to institutions with global reach, hundreds of employees, and multimillion-dollar budgets. They have diverse reach and influence. If you have seen one think tank, you have seen one think tank.

One way to assess think tanks is to consider their mission and sources of funding. Both shape what they do and how they do it. For instance, some think tanks focus on a single mission, like environmental issues. Others, like the Brookings Institution, cover a wide range of domestic and foreign policy areas. Funding matters as well. In some cases, research is funding dependent. If there isn't a funder to pay for the staff and operations for the research, there is no research. All that said, in the end the work of a think tank is best judged by the quality of the staff, their research, and the impact of their operational activities. Most public-facing think tanks are transparent about their governance, funding, and the research they produce and the activities they conduct. Judge for yourself.

This community of research institutions has their greatest impact when they network and leverage their contributions to public policy development and advocacy. A few illustrative examples have already been noted in *Digital Dominance*. For instance, our collaborative think tank project on reforming the structure of the Department of Homeland Security not only benefited from the many talents we drew on but also the broad-based non-partisan agenda of the team which demonstrated that our effort was intended to advance good public policies, not just one party's political program.

Another example noted was the China Transparency project discussed in the last chapter. Not only did this effort show the collected intellectual and analytical capability that could be turned on a problem, but the think tanks also demonstrated they could undertake efforts governments had not or could not. No liberal government has produced a comparable, unclassified accumulation of data.

Lisa Curtis, who was mentioned earlier in the chapter and in the prologue, is another case in point. If you do an internet search, you won't find a fistful of research papers published by Lisa. She is not a writer. That is not atypical in the think tank world. The stereotype that the great analysts just sit in their book-lined offices thinking and scribbling great thoughts is wrong. Although writing papers is fine, it is what think tankers do with their ideas that really make a difference, and that difference often materializes through purposeful network action.

Arguably, not all the institutions in the public policy space are forces for good. Some suffer from poor governance. Some drift into irresponsible political action. The Southern Poverty Law Center (SPLC) offers an exemplar of both.

Established in Montgomery, Alabama, in 1971, the center played a prominent role for bringing legal cases against white supremacists and other hate groups. In the contemporary public sphere, the center continues to play a prominent role. Amazon, for instance, uses the center to vet charitable organizations that participate in its "Smile" program to weed out extremist groups. Today, however, the center's reputation is significantly damaged. According to one assessment, "since its victories over the Klan in the 1980s, the SPLC has been widely criticized by both right-of-center and left-of-center observers for its excessive fundraising and controversial methodologies."[13] In recent years, the center was rocked by "allegations of a toxic workplace culture that discriminates against women and people of color."[14] One nonprofit organization claimed they were specifically targeted for armed assault because the center had labeled them a "hate group," based not on facts but as part of an opposition political campaign.[15] The SPLC wrote a damning report accusing anti-Islamist organizations combating extremism and transnational terrorism as a hate group. That earned the center a lawsuit that they lost, costing them millions of dollars.[16] Whether the center recovers its reputation and influence remains to be seen.

Although all think tanks are not equal or equally credible as a force in the modern networking world, collectively they are a powerhouse of ideas and action that cannot be ignored. That think tanks can achieve such a significant impact in shaping civil society and government also makes them targets for influence by states and other nonstate actors. Some of these efforts are malicious, intended to disrupt or undermine their operations. Previously, for instance, *Digital Dominance* highlighted that think tanks are often the target of cyberattacks. Like other groups that compete in the networking space, think tanks have to take appropriate measures and countermeasures to protect their operations. It is part of the price of doing the public's business in a public space.

Other efforts to affect think tanks are by influencing governance or through directing funding. One contentious issue in the United States, for example, is concern over how foreign government grants may influence the work of some think tanks.[17] Responsible governance, transparency, and ethical practices are crucial for think tanks to retain trust and credibility in the public sphere (as the case of the SPLC reminds us).

Angels on High

In addition to think tanks, there are other NGOs that can impact matters related to national security. This includes a variety of NGOs conducting activities from humanitarian relief to developmental assistance. Some of these NGOs operate as neutral organizations. They specifically eschew aligning with parties in a conflict or, indeed, in conducting any kind of operations in a country. Another category is the non-neutral NGO, like Spirit of America. The role of these organizations may be even more impactful in great power competition because they are more likely to amplify state power and support national interests.

As noted before, traditional humanitarian groups hold that NGOs operating in conflict zones should stay scrupulously independent.[18] The Red Cross established this concept in 1921 as one of its core principles. The United Nations later insisted that all of its humanitarian missions would be characterized by neutrality, impartiality, and operational independence. The idea is that aid should never be distributed with political, economic, or military objectives in mind. "These principles are not primarily moral values, but rather a means to secure access to those who suffer the brunt of conflict and violence and to enhance the effectiveness of aid," said Angelo Gnaedinger, director-general of the International Committee of the Red Cross, in 2007.

The non-neutral NGO model followed by Spirit of America and others is intentionally different. "We advance human security and well-being," says Jim Hake, the head of Spirit of America, "but taking a side breaks new ground in international assistance." In contrast, Jenny McAvoy of the humanitarian coalition InterAction argues that non-neutral charitable work can have undesirable effects in conflict zones: "punishment of vulnerable people, widespread suspicion of humanitarian organizations, denial of access to affected populations, targeted attacks on humanitarian workers, manipulation and diversion of aid to serve political goals." Alas, all those issues already bedevil even the most "neutral" aid efforts today. From ISIS to the Lord's Resistance Army, militant groups refuse to tolerate even adamantly "impartial" humanitarian intervention.

Not everyone, however, thinks that impartiality is really that valuable. Nadia Schadlow, when she was at the Smith Richardson Foundation (one of Spirit of America's supporters), argued that the space for "neutral" aid is relatively narrow. "Once any long-term effort to alleviate suffering begins, it becomes political," she points out. So "not taking sides" is often more a theory than a practical reality. For instance, argues Jim Hake, traditional humanitarian boundaries can't safeguard relief workers today. He cites the beheadings of aid workers by ISIS as an example. "Terrorists put us all at risk. By helping our troops defeat them, Spirit of America makes it safer for neutral organizations to do their work," he asserts.

Spirit of America isn't the only private organization that takes the United States' side in conflict zones around the world. America Abroad Media is a nonprofit supported by numerous donors like the Carnegie Corporation and the Stuart, Diana Davis Spencer, and Starr Foundations. America Media Abroad produces for broadcasting partners in conflict zones programming that encourages free inquiry and alternatives to extremism. The 501(c)(3) National Strategy Information Center promotes democratic processes in poor countries. For instance, its anti-corruption programs teach practical ways of reducing bribery of police officers and soldiers, treating citizens respectfully, and building community bonds.

Another good example of angels in action is the Center for International Private Enterprise (CIPE). In the prologue, I mentioned it was an official from CIPE trying to help their friends in Afghanistan who first got me involved in the evacuation effort. CIPE, largely funded by the US government, organizes programs in several countries important to US interests, helping build democratic institutions that enable environment for business and entrepreneurship. As a case in point, *Digital Dominance* previously mentioned the example of Moldova struggling to escape the influence of Russia's shadow. CIPE has been running programs in the country for some time. They have helped bolster the anti-corruption movement of the new government. CIPE, for example, organized training for forty officials from seventeen different national agencies responsible for implementing a new law on transparency in governance.[19] They also conduct many other initiatives on free-market practices and good governance in countries around the world in the developing world that are strategically relevant to the United States.

These organizations have the most impact when their interests align and are integrated into cooperative campaigns that support national security–related objectives. The US government, as example, developed several noteworthy programs related to supporting women in society. One example is the Women's Development and Prosperity Initiative. Another is the US government support initiative for Women, Peace, and Security. Both of these efforts are operationalized through a coalition of NGOs, some funded by the US government. Others operate in concert with these efforts.[20]

Of course, not all angels have shiny halos. Some organizations that purport to advance constructive and positive agendas beneficial to civil society and public policy may be masking darker intent. One case in point is the Council on American Islamic Relations (CAIR). CAIR purportedly acts as an advocate for civil liberties and religious freedom. Nevertheless, CAIR has been cited for its links to extremist groups. One assessment noted, "Over the years, CAIR's alleged ties to Hamas have proved troublesome for the organization. In 2007, federal prosecutors reportedly designated CAIR a co-conspirator with the Holy Land Foundation, a group that was eventually convicted for financing terrorism. In 2014, CAIR was designated a terrorist organization by the United Arab Emirates and multiple CAIR members have been arrested on charges related to terrorism."[21] CAIR also has governance issues. In 2021, for instance, leaders in one chapter were accused of harassment and misconduct.[22]

In practice, NGOs, like think tanks, represent many different groups and offer many different capabilities and a wide variety of goods, services, and advocacy. Identifying the ones that can play a constructive role in networking and eschewing or countering those that are a menace to liberal states is part of the challenge of fighting for an edge online and offline.

Strength for the Fight

Social networking warfare is empowering for nonstate actors, particularly in states where the public sphere is not micromanaged by authoritarian governments. Even under the shadow of adversarial regimes, NGOs are not powerless, evident in the extraordinary efforts that states like China, Russia, and Iran take to monitor, manage, or suppress their activities. Conversely, recognizing the power and influence of NGOs, adversarial states make strategic efforts to exploit their efforts for their own interests. Some of these initiatives manifest themselves in the support or use of armed groups.

The role of nonstate groups reminds us that social networking competition in the modern world transcends government action and the public and private spheres. Networks have the capacity for influence and action across the spectrum of human activity. They are, and will remain, part of the business of doing national security.

What is also true is that the nongovernmental "business" will be part of the business of doing national security—and that just doesn't go for companies that are government contractors. The largest global corporations in the "big tech" world have an especially important place in great power competition. That is the subject of the next chapter.

13

GUNS AND BUTTER

Andy Puzder is bad for businesses that are bad for business.

There is no more a zealous defender of free markets and the free enterprise system than Andrew Franklin Puzder.

Andy was born in Cleveland, Ohio, a boy about as blue collar as the town itself. He worked his way through college doing construction, landscaping, and house painting. Then he quit college to play the guitar and perform in local bands. That led to a job in a guitar store and a career in business.

He later finished his degree and got another one from law school. For years, he practiced law, including representing a local fast-food chain in St. Louis. He finished as the head of CKE Restaurants, turning Hardee's and Carl's Jr. into a global chain of restaurants. When he was done, the company had over 3,800 locations.

Puzder also became a prominent proponent for causes. Chief among his passions is being an advocate for free enterprise. No longer worried about needing a day job, he is a frequent lecturer on economics and politics. He is also a published author and established media presence, as well as an adviser on boards and a contributor to think tanks.

What occupies his time today is pressing businesses to be good for business rather than allowing their cooperate boards, unions, and employees to drive company policies based on politics. He is particularly an advocate for fighting environmental, social, and corporate governance (ESG), which contends that these nonfinancial factors should drive corporate decisions. Puzder argues that is not only bad for business, but ESG is bad for employees, consumers, and everyday citizens. ESG is often not even about achieving better societal outcomes. The standards are often imposed to advance political agendas. His advice is to "focus on profit, not politics."[1] Companies better serve their customers, communities, and country when they focus on what companies do best—act like businesses.

Over time, Andy's expertise and judgment gained national attention. In 2025, he was nominated to be the US ambassador to the European Union.

What Andy is fighting for has big implications for national security and networking warfare. Here is why. The biggest technology companies, which also dominate social networking services online, play an important role in the national security of liberal states.

The Military-Industrial Complex

Dwight Eisenhower may have been one of my favorite military planners and presidents, but he didn't get everything right. He was wrong about the military-industrial complex taking over America (though he was right to worry). If he was right, there would never have been successful, wild-eyed entrepreneurs like Andy Puzder.

In his final speech as president, Eisenhower warned, "In the councils of government we must guard against the acquisition of unwarranted influence, whether sought or unsought, by the military-industrial complex." He believed excessive government spending on national security would lead the country being run by to a cabal of defense industries, the Pentagon, and big government who would make decisions that suited them best, undermining democracy, and sucking up the wealth of the American people.

To be fair to Ike, Ike had something to fear when he declared, "The potential for the disastrous rise of misplaced power exists and will persist." During Eisenhower's presidency, the government peacetime spending to fight the Cold War was unprecedented. Fifty percent of the federal budget went to defense. Most of the research and development spending in the United States came from the government. Defense companies were some of the biggest corporations in America. Even companies that didn't build planes, ships, and tanks played a role in national security. AT&T, for example, enjoyed a government-sanctioned monopoly in telephony. The company owned Bell Labs, the most powerful scientific and engineering private sector place on the planet. Bell Labs did all kinds of national security–related work.

As historian Martin Medhurst notes, "Eisenhower was deeply concerned about the growth of the federal government and the systematic loss of state and local autonomy. He was concerned about a government that spent more than it took in, a government in which the twin threats of spiraling defense spending and an ever-larger federal largess threatened to turn the country into a "garrison state where individual liberties might be easily lost."[2] Ike wanted to make sure the United States had a private sector that remained vibrant, growing, prosperous, and independent. He considered that just as important for national security as armies and navies. To win the Cold War, the United States needed guns and butter, not one or the other.

Jumping ahead, the military-industrial complex did not take over the United States. Princeton scholar Aaron Friedberg wrote a fantastic book, *In the Shadow of the Garrison State: America's Anti-Statism and Its Cold War Grand Strategy* (2000), which explains how free enterprise survived the Cold War. When the Cold War ended, it was self-evident—except for critics like Howard Zinn, William Appleman Williams, and Noam Chomsky—that America was still a democratic republic.

The last thing Americans must worry about today is being ruled by the military-industrial complex. In the 1990s, the US economy began a radical transformation. Today, no defense company is in the top tier of US business enterprises. Their place was crowded out by "big tech," technology companies like Alphabet Inc. (Google), Amazon, Facebook (now Meta), Apple, and Microsoft, which dominate the American business space the way industrial giants like DuPont and American steel and railroad companies did in the nineteenth and twentieth centuries. These contemporary giants of the economy have the capacity to shape how everyday people live their lives. And unlike the phantom military-industrial complex, that might be a worry worth worrying about.

The shift in the US private sector landscapes matters for national security and networking in different ways. Let's walk through the list of concerns; there are five.

Global Reach. Big tech has global reach and a global customer base. Ninety percent of Facebook's users (as has been noted before) are outside the United States. That raises a couple of concerns. Will companies like Facebook put the interests of other nations ahead of the United States? Will companies, to do business in other countries, do the bidding of these countries against US interests? These pressures are real. Google, for instance, confronted ethical challenges in providing search operations in China in the face of the restrictions of the Great Firewall.[3]

Going Woke. Woke has become the derisive term of art for advocacy of social, economic, and cultural causes driven by progressive partisan political agendas. An example, as previously noted, is the case of Google, which abandoned defense work over employee protests; its employees interpreted doing defense work as a violation of the company's business pledge to "do no harm."[4] These controversies are also reflected in Andy Puzder's concerns with the ESG movement. To be fair, the politization of business practices is not only a charge leveled by conservatives. Liberal groups have also argued they are unfairly targeted.

Wokeness is not only a potential domestic concern that can sow discord and discrimination. Woke as a weapon is already being weaponized by adversarial states. Vivek Ramaswamy (an advocate for fighting ESG before he became a presidential aspirant), for example, points out the case of US universities adopting "trigger warnings" before being exposed to sensitive ideas "on social questions like race, gender, and climate change. It turns out China took note of the effectiveness of the woke movement . . . based on a new Chinese law, top universities including Harvard and Princeton have begun to label certain courses . . . if they teach any material that China may consider offensive."[5] The potential for the abuse of this practice is obvious and might just as likely be used to influence corporate behavior. What might well be even more injurious is companies self-censoring out of fear of antagonizing China or others and then using the cloak of acting woke as a defense of their actions.

Privacy Piracy. Tech companies have become massive repositories of information on their users. The concern is that they can gather so much data and have such powerful instruments to manipulate that data that the companies will obliterate individual privacy. This is a threat that could be exploited by the companies themselves, host governments, or foreign entities. TikTok, the video-sharing app owned by a Chinese company, has, for example, been flagged as representing a significant concern.[6]

Celebrating Censorship. While big tech often removes content for public safety purposes, companies also act to combat what is often labeled "misinformation," which includes posting articles, videos, and podcasts intended to deceive the public. Voices on both left and right have accused tech companies of abusing this practice to impose unfair censorship, an abuse that could be exploited by companies or governments. "Social media companies were ostensibly created to democratize access to information and give ordinary citizens a voice in the public square," argues my associate Kara Frederick. "Today, reality has proven much different."[7] Censorship is a real concern. "Security tools," she warns, "are increasingly repurposed to police viewpoints that run against the progressive leftist narrative. Tech companies already on track to weaponize their counterterrorism tools in this way." Marrying up the capabilities of big tech with political censorship could literally kill democracy.

Here is a small example. The think tank I work for, for instance, complained when an ad for a book by one of our scholars was banned because "content that revolves around controversial or highly debated social topics is not permitted." After complaining that this was an absurd standard that "wouldn't allow ads for the biggest bestseller in history—the Bible—a book that stirs incredible debate and is considered controversial by those who don't believe it," Amazon reversed its ruling.[8] The incident reflected the concerns that tech censorship is arbitrary or politically motivated. The reality is, big tech isn't any better an arbitrator of truth than the government or anybody else.

Crushing Competition. Big tech dominates the tech stack which allows other companies to operate. These companies also have a commanding share of goods and services across the tech sector. This represents power that some argue could be used to an unfair competitive advantage. Already cited, for instance, was the case of Parler. For several weeks, the app was forced offline because companies refused to provide supporting tech services.

Together, these issues raise concerns about potentially consequential impacts on national security from undermining innovation and competitiveness to diminishing the resilience of civil society and compromising or undermining the capacity of the nation to defend itself from foreign threats or protect national interests.

These concerns are not fearmongering. Issues that raise red flags pop up all the time. Here are some examples. In 2021, the *New York Times*, based on a review of requests for bids for contracts by the Chinese government, concludes, "Officials tap private businesses to generate content on demand, draw followers, track critics and provide other services for information campaigns. That operation increasingly plays out on international platforms like Facebook and Twitter."[9] The report noted some efforts by social media platforms to address this abusive behavior. For instance, Facebook deplatformed over five hundred accounts used to spread articles by a fictitious Swiss biologist criticizing US "interference" in the World Health Organization and global COVID-19 response, after determining the accounts were mimicking reports from Chinese state media.[10] Yet, the tremendous scope of the entire effort and the extent to which the government contracted with media companies to conduct these operations raises questions about the probity of their dealings with the regime.

Another case that drew attention was a long-standing arrangement between the Chinese government and Amazon that facilitated distribution of official government writings. Among the accusations was that Amazon had to agree to block negative comments on a collection of President Xi Jinping's speeches and writings.[11]

Hubris

There is, of course, a vociferous debate over how serious these concerns are and whether remedies might pose just as serious a problem for the freedom, prosperity, and safety of citizens, communities, and countries. For instance, in the United States there are arguments for breaking up big tech companies through antitrust legislation. These are countered by anxieties that this might undermine national competitiveness or give the government too much power over determining how the private sector should function.[12] The debate is as visceral as the argument over the threat of contractors in combat discussed in the last chapter, Frankly, this dispute is a lot more consequential for the future of the free world's ability to compete, survive, and thrive.

Why worry? How this all plays out will largely depend on the role big tech decides to play in great power competition. Here is the thinking. For starters, the reason why the private sector has become more important in national security is simple. The private sector is more powerful. There is not much that states either liberal or authoritarian ones can do about that. If states exercise too much control, businesses will lose the competitive practices that make them such powerful competitors. Out goes innovation, risk-taking, and initiative. In comes central planning, picking winners and losers, and corporatist behavior that won't deliver world-class economic power.

Let's be honest, the importance of the global private sector even holds true for China. China became rich through a twenty-year period of economic liberalization. Beijing continues to power its growth from doing business with consumers in the free world, intellectual property theft, corruption, and shady business practices, all tactics that would not flourish without a world of free enterprise to exploit. Without a free market, China is an economic heroin addict, hoping it never runs out of heroin because as soon it does, the Chinese economy will go through a painful withdrawal. China's growth cannot be sustained through domestic markets and centralized economic planning.

It is also true that some poor authoritarian states can generate some significant national power. North Korea built an impressive nuclear weapons program with a society that can't feed its own people. Iran pays surrogates to attack its enemies. Russia has proven itself a reckless military threat. Even destitute Venezuela can afford to keep its ruling class in steaks and fine cabernet. This is, however, not that hard. In an authoritarian regime, no matter how meager the resources, leaders can direct cash to make sure the things leaders care about get taken care of. They cannot, however, do what a modern private sector free enterprise can do. Take Russia. For as scary as Vladimir Putin seems sometimes, it is worth remembering that for all his country's vast land and resources, the entire economy of Russia is smaller than that of Texas.

So here is the bottom line: If the private sector stays private, in the end what and how well the private sector does will depend on the private sector. That's why big tech is a problem.

The problem with big tech is best illustrated with a true story. Several chapters back, I recalled when I had been asked to accompany foreign military attachés on a tour of Silicon Valley, which remains the epicenter of the big tech industry. One of the briefings we received at one of the companies included a tutorial on geopolitics from a tech perspective. Here I paraphrase, but this how I remember the lecture going. "I don't think we really need the United States," the discourse started. "Silicon Valley is wealthier than most countries on earth. We could get along on our own. Sometimes, I think we would be better off." I was a bit shocked by the hubris. Last time I checked, big tech didn't generate its own electricity. It didn't build its own highways. It didn't guard its borders. It didn't protect intellectual property rights. It didn't provide courts so big tech companies can sue big tech companies (at one point, Google and Microsoft spent more money on lawsuits than research and development). Big tech doesn't do any of the things nations provide so that business can go about the business of doing business. In the United States, as rich as big tech is worth (Apple in 2021, for example, was worth about $2.1 trillion), tech could not cover the cost of running government, let alone cover the national debt. So, honestly, give me a break on the whole idea that big tech doesn't need nations.

The tale of a big tech blowhard is worth retelling because it reminds, in the end, companies are run by people, and people can have some silly ideas. I am not suggesting the yahoo with an inflated image of his place in the world speaks for every board member of every big tech board, but it does recall an adage: "Sometimes the stupidest people are smart people." Because they are smart about some things, it doesn't mean they are smart about everything. There is a shocking lack of humility in Silicon Valley. This is not to suggest that we should tell companies how to run their companies, but we would all be better off if they stuck to doing that, rather than dictating to the rest of us how the rest of the world should work.

What's more, it is not at all clear that even though big tech does business around the globe that the big tech world understands how the world turns and their place in it all that well. In 2021, for instance, the head of Apple declared, companies had a responsibility to do trade around the world to promote "world peace through world trade," while he had to acknowledge that there are different laws in other markets.[13] This statement suggests that even the head of one of the biggest big techs doesn't grasp how commerce works.

Getting wrong the role of how the free market works in the world is nothing new, even for smart people.[14] Sir Ralph Norman Angell, for instance, thought he had all the answers. In 1909, he published *Europe's Optical Illusion*, a pamphlet arguing that the global integration of economies through trade and industrialization made total war obsolete. The outbreak of World War I dented this theory somewhat. But such was the power of Angell's argument—combined with the perspective of European elites—that he received the Nobel Peace Prize in 1933. World War II started six years later.

Angell and others missed a key point. It's not just the intensity of trade between nations that influences the tide of war and peace; the kind of nations engaged in trade matters as well. The old British adage "Trade follows the flag" should have had an important postscript: "War often follows trade." When free market nations rubbed up against mercantilist and other "not free" economies, the friction often produced bloodshed.

It is only economic activity between nations that share a commitment to economic freedom where commerce tends to flow peacefully, without rancor. Indeed, strong trade ties between free market nations tend to promote national security. The reason for that is economic freedom helps nations generate wealth that allows them to defend themselves and creates a community of nations with a shared interest: protecting their right to freely exchange goods, peoples, services, and ideas. This common bond promotes the cause of peace by creating strong, self-reliant, sovereign, and independent nations interested in preserving the mutual freedoms, which in turn allows them to engage commercially and prosper.

Economic freedom hinges on an institutional framework that facilitates all individuals exercising their liberties in the marketplace. In addition to accommodating free trade, that structure includes institutional commitments to fight corruption, protect property rights and the sanctity of contracts, and pursue responsible fiscal policies.

In short, commerce, despite what the head of Apple thinks, doesn't spread freedom. The last development liberal states should expect is that when they freely trade with authoritarian states, those nations will catch liberty like a cold.

Responsible Behavior

The irony is, if states had sound public policies and companies assessed their risks of doing business in authoritarian states like China correctly, then competing globally without compromising the freedom, security, and prosperity of the free world would be manageable. Competing with China, for instance, could be way better managed.[15] Here are four actions where the public and private sectors could join forces to deal with the challenges of trading with China.

Push Transparency. The more the free world makes transparent to the world the Chinese Communist Party's (CCP's) mendacious actions, the more folks will rethink how and when they do business with China. There are plenty of other actions that ought to be exposed more thoroughly, from predatory lending to corruption, disinformation, meddling in other countries, intellectual property theft, and widespread human rights abuses, to name only a few. These are the kinds of activities the China Transparency project is trying to uncover. The more people know, the more risk-informed decisions they will make in dealing with China. That will take the excuse of "we just didn't know" off the table.

Make It Easier to Do Business at Home. Don't ask, "How can we force people to stop doing business in China?" Ask, "What obstacles can we remove to make it easier for people to do business here?" In the United States, for example, look at the federal, state, and local levels and ask what red tape and other impediments to doing business we can trash. Telemedicine, for instance, is a good example of where some states are striking out and delivering cutting-edge innovation that is making a dramatic impact. Be more proactive. From advanced manufacturing (using innovative technology to improve products or processes often without additional labor and at a lower cost) to the gig economy (short-term and freelance work replacing or supplementing permanent employment), let's make it easier to put people to work.

Trade with Our Friends. China is not the only trade and supply chain option. Let's figure out how to do more business with our allies. It is ridiculous, for example, that the United States and Europe can't partner on viable free market options for 5G telecom networks. We shouldn't have to buy from the telecom company Huawei, a tool of the CCP, and accept all the security risks that that brings.

These are steps that governments, the private sector, and civil society can take right now. They would help, not hinder, our efforts to grow the free world's economy. They are practical, suitable, and feasible. Most of all, they would signal to the CCP that the free world is nobody's patsy anymore.

Just Do It. If the West is going to exploit technology for national security, it is going to have to buy a lot of technology. It presently is not. Overall, gross domestic product spending on defense looks flat compared to countries like China and Russia, who are investing a lot in a lot of capability. Accounting for the advantages countries like China have in relative purchasing power, what they are spending on offensive military capabilities in comparison to the West is stunning.[16] Beijing can just demand what it wants and dictate the price. The private sector in the West is not going to deliver national security capabilities unless there is a market. One complaint from the market is that the demand signal for defense goods and services is weak compared to needs.[17] Another problem is the petty squabbling between free nations that are more concerned about preserving market share for their own companies than arming against the coming storm. European nations, for instance, formed the Permanent Structured Cooperation (PESCO) to facilitate building out defense capabilities, but the project specifically excludes defense cooperation with the United States and Canada. This, in turn, prompted complaints from US officials that PESCO looked more like fostering protectionism than buying protection. In turn, the Europeans object to "Buy America" provisions in US laws which, they point out, exclude European companies.[18] If governments don't start buying what they need, there won't be an industrial and services base there to provide what they need.

Course Correction

The problem comes when Silicon Valley thinks what is good for Silicon Valley is good for the rest of the world. Add to that pathology the problem of playing politics with the private sector (the kind of ESG issues that Andy Puzder is always complaining about) and the free world has a bigger problem than just greedy companies.

When private sector power is being harnessed for political interests, that's the point when companies start to think their self-interests have special virtue, as opposed to just being what they are, self-validating political opinions. This is a toxic cocktail that is unhealthy for civil society or good governance. The resulting hangover can manifest itself in bad ways. Here are four.

Corporatism. This is a political structure where corporations become part of the political institutions of the state; the interests of the regime and the companies comingle. The Chinese concept of civil-military fusion is a form of corporatism, as was the military-industrial complex that President Eisenhower warned against.

Regulatory Rule. This occurs when corporations prefer or advocate for regulations that govern the behavior of industrial or business sectors. When companies invite regulation, they often so do with the intent of shaping the regulations to ensure they have a business or market advantage that reduces competition. These rules can often have a deleterious effect. For example, the Merchant Marine Act of 1920, the Jones Acts, was intended to protect the US merchant marine industry. In practice, it made American industry less competitive, but the law is still zealously defended by a handful compa-nies because it preserves their business with customers that have no other option.[19]

Rent-Seeking. This is a term used to describe a business that seeks to in-crease wealth without any increase in productivity.[20] A common form of rent-seeking is when oligarchs who control business sectors in authoritarian countries are granted monopolies or exclusive or noncompetitive contracts by the government. This is a common practice in Putin's Russia.

Picking Winners and Losers. When government intrudes in the practice of free enterprise to empower some companies, technologies, or business sectors at the expense of others, it risks undermining the dynamism of the marketplace and usurping the choices of consumers to decide for them-selves. Consumers likely would never have the many options, goods, and services they enjoy today if AT&T had retained a monopoly over telephony. In part, the dynamism of the digital economy in the United States emerged because the breakup of AT&T coincided with the rise of the personal com-puter industry.

All of these pathologies result in unhealthy practices that undermine the relationship between public and private sectors in liberal states where gov-ernments either wind up colluding with business, controlling private com-panies, or are unduly influenced by commercial interests. When the power of the companies is as massive as the tech center is in the free world, the potential for undermining the freedom, security, and prosperity of citizens is very real. The greatest potential for abuse is when the politics of compa-nies align with the interests of a political party and that party holds sway over the instruments of government. Then the attraction of mutual interest between the public and private sectors becomes dangerous to the practice of free enterprise.

Preserving Economic Freedom

The private sector can best serve the public interest when two conditions prevail. The first is what Andy Puzder calls "getting back to neutral," get-ting companies off their addiction to ESG. Vivek Ramaswamy, another suc-cessful entrepreneur turned evangelist for protecting free market competi-tion (turned politician), makes a strong case for back to neutral in *Woke Inc: Inside Corporate America's Social Justice Scam* (2021). Ramaswamy agrees with Puzder. Wokeness, he argues in an interview in the *New York Post*, "is remaking American capitalism in its own image. . . . Today's captains of in-dustry do it by promoting progressive social values. Their tactics are far more dangerous for America than those of the older robber barons."[21] This is not to argue that people in companies should not have political views or that one person's politics are better than another's. What Puzder and Ramaswamy warn about is the corrupting influence of weaponizing politics as a business strategy, using the power of the private sector as a heavy-handed tool to reshape civil society and drive specific political outcomes. Companies should stop being sledgehammers for ESG.

The second condition is that big tech ought to be patriotic. This does not mean that businesses should align with the government. What companies should do is align with the national interests of the liberal states that are their home. Here is why.

Liberal states are nations of law. That just doesn't mean they are states with laws. All states have laws. China, Russia, and Iran have lots of laws. Liberal states practice ordered liberty or the rule of law. Their laws are qualified by two general conditions. One is that there is equality of treatment under law. No one, for instance, is exempt from laws (like Russian oligarchs). The second condition is that societies refrain from establishing laws that infringe on human rights or inordinately impinge on individual liberty. No laws, for example, should legalize human trafficking. Free market enterprise thrives in these societies. In consequence, corporate institutions as individual citizens ought to seek perpetuating these societies. That is the one true corporate responsibility based on self-interest. These are interests that transcend political parties. They are the definitional concepts of a liberal nation. Companies can't do business in the free world if there isn't one.

There is, of course, a long history of companies beyond just the defense manufacturing base supporting the national security needs of their nation. Public-private sector cooperation in the United States during World War II is the penultimate example.[22] The consequence of getting the role of big business right in the era of great power competition is just as serious for the fate of free nations.

Agenda

Winning in networking warfare for either liberal states or nonstate actors in the liberal world might not be possible without a powerful, resilient, innovative big tech collaborative space. Left to its own devices, big tech probably won't deliver. Here is why. After tracking the industry for over a decade, there is no sign that it has figured out its collective role.

There are signs, of course, that some companies sometimes make the right decision. One example, cited before, is the case in 2021 when Microsoft, which owns the networking platform LinkedIn, decided to discontinue operating the platform in China. The company knew that the Chinese government was exploiting the service to spy on their own people. Microsoft could not allow that behavior to continue without compromising the company's reputation, so they rightly pulled the plug. That said, there are more than as many examples of companies getting it wrong. Leaders and employees of many these companies still just don't get it. They don't acknowledge that the free world is for the fight of its life and that woke behaviors are making liberal states less, not more, competitive.

The behaviors of free market giants is best changed in the same ways the excesses of the nineteenth-century robber barons were tamed from a combination of appropriate government and consumer actions. Let's outline those here.

Legislating. There is an appropriate place for legislating activities that curb abusive behavior. What is critical is that any reforms be rooted in the protection of free speech. In the United States, for example, the First Amendment to the Constitution should be foundational to any reform effort.

One frequent target in the United States for legislating is the Section 230 provision of the Communications Decency Act, which holds "no provider or user of an interactive computer service shall be treated as the publisher or speaker of any information provided by another information content provider." This measure was intended to prevent endless lawsuits against platforms for being sued by the third-party content this carried. As big tech plays a greater role in policing, monitoring, educating, and mediating the content on their platforms, some raise concerns that they are like publishers but remain immune from any abusive practices other publishers are subject to under law. This has triggered a running debate over ending or mending the legislation. There is a good case to be made for refining the law.[23] Getting the laws right for the United States and other liberal nations will be a key factor in determining how constructive a role big tech plays in future competitions.

Innovating. Nations like the United States, which have a federated system of government, have the benefit that states and provinces can serve an incubators of innovation and good governance. Kara Frederick argues for a "simultaneous, multipronged approach to securing freedom of expression in the digital world includes promoting the principles of federalism through constitutional state legislative action, efforts by technologists to build platforms where freedom of expression is protected, as well as vivifying civil society efforts to promote transparency within these companies."[24] Every free society needs suitable checks on the authoritarian tendencies of central governments.

In the United States, several states are introducing methods to address alleged big tech abuses. In 2021, for example, in Florida, the state government initiated an effort curb online censorship.[25] The success and failure of these efforts may offer examples for others.

Transparency. Civil society has an important role in protecting its place online from abusive activities by big tech. Think tanks and other nongovernmental organizations can turn their efforts on bringing clarity to big tech corporate behavior. The American Principles Project, for example, tracks donations from big tech to organizations influencing tech policy.[26]

Consumer and Shareholder Activism. Both have a powerful voice in impacting corporate behavior if they elect to use it. The Alliance for Defending Freedom, for instance, helps shareholders raise challenges to irresponsible corporate policies including disputing Amazon's partnership with the discredited Southern Poverty Law Center.[27]

Building Out the Tech Stack. As described previously, new companies can't get online, innovate, and offer new goods and services without a reliable, resilient tech stack. Our think tank, for example, requires 33 technological products and 128 technological services just to support day-to-day operations. As much as social networking is really a human thing, you cannot do it online without tech. There are already service providers of platforms and products, like Gettr (a social media microblogging site) and Conservative Stack (a content management platform), that ensure customers they will be protected against political censorship.

Buying Stuff. The US government needs to figure out how to become a better consumer of private sector goods and services. This complaint has been around since we started having a government. On the one hand, nobody wants to waste taxpayer dollars or forge the cozy government-business relationships that add to fraud, waste, and corruption. On the other hand, if government wants the private sector to deliver cutting-edge national security products efficiently and economically, it is going to have to be a better customer—slashing red tape, unrealistic and overly prescriptive requirements—and figure out how to buy the stuff it needs when it needs it. This is not just a US problem. Collectively, the free world needs to recognize it has the fight of its life on its hands. If the free world doesn't start arming up for global competition in a serious way, nations will soon find that the private sector is not prepared to ramp up for what is coming.

Education. We need more savvy national security networkers pretty much everywhere, but no sector of society needs them more than big tech. That's not to say that big tech companies don't hire people with national security experience—they do. They hire veterans. They hire intelligence and cybersecurity experts. These voices are not all that influential in the corporate policy circles, nor does big tech make a serious effort to groom leaders with the skills, knowledge, and attributes that would enable them to play a constructive role. There is no educational structure adequately addressing this shortfall. This shortfall will continue to plague big tech.

None of these initiatives are silver bullets for ensuring big tech will be the partner the free world needs for great power competition. Collectively, however, they represent a powerful force for fixing what has become a huge vulnerability in how the free world competes to protect freedom.

Wild Card

The lineup of companies dominating the world of big tech today are not the same as they were over a decade ago when *Wiki at War* hit the bookshelves, audiobooks, and Kindles everywhere. The lineup ten years from now could look even more different, as could the face of civil society and the character of governance in states around the world. Those changes could well be impacted by emerging technologies. The next chapter looks at the wild cards that are most likely to shape the future social networking war.

14 EMERGENT

J. V. Venable does not look like he would fit in the cockpit of a fighter jet.

From birth, John "J. V." Venable was a long, lanky character. By the time he graduated from Ohio University (where Pete Mansoor teaches today), the distinguished graduate of Air Force ROTC was basketball player tall. It looked like they would have to press and fold him to fit in the seat of a fighter aircraft.

JV, like John Boyd, always thought he was built for speed. In a twenty-five-year air force career, he did plenty of flying in peace and war, serving at sixteen—count them—sixteen locations around the world as a forward air controller, fighter pilot, staff officer, and commander.

His weapon of choice was the F-16 fighter, a plane he has flown throughout the United States, Europe, the Pacific, and the Middle East. I once asked how he fit into the cockpit. He confessed, not well. "The way you proved you were not too tall was there was supposed to be a fist's distance between the top of your helmet and the canopy. The reason for that was that the canopy was pliable. If a bird hit it, it would give, and the impact might knock out the pilot." JV held out a flat hand. That was how much head space he had. I guess his plan was just to try not to hit any birds.

When JV wore a uniform, he really was kind of a John Boyd character. He was a doer and a thinker. Once as the weapons officer for an experiment at Pope Air Force Base, North Carolina, he developed a suite of "killer" tactics, earning a nomination for the prestigious Claire Chennault Award.

He had three combat tours (more than John Boyd). In 2004, he may have run into some of Zabihullah Mujahid's friends in Afghanistan where JV led the largest combat operation in the air force anywhere in the world:

16 squadrons and 1,100 men and women, supporting ground and air operations in both Afghanistan and Iraq.

How good was he? Good enough to be tapped as commander of the celebrated Thunderbirds, the elite flying demonstration squadron of the US Air Force.

In addition to being a real-life *Top Gun* (1986) character, JV had two other world-class skills. He knows a thing or two about inspiring people, authoring a book, *Breaking the Trust Barrier* (2016), on managing high-performing teams like the Thunderbirds. He is a sought-after speaker on leadership.

JV is also a bit of a mad scientist. He holds a master's degree in aeronautical sciences from Embry-Riddle Aeronautical University. JV also has a degree in strategic studies from the Air War College.

After retiring, he served as vice president of a research and development company specializing in explosives detection. Leading an army of doctorates who understood physics but not economics and a business team that understood the bottom line but not quantum mechanics and chemistry was no mean feat.

I can't remember exactly how I wound up finding and hiring JV, but it was a genius move if I do say so myself. Any smart person can master science. Any thoughtful person can figure out how to make things work in the real world. A patient teacher can explain how complicated things work to others with plain common sense. Finding one person who can do all of that is like finding a bank that can print its own money and rob itself.

I never asked JV why he dumped the world of research and development for a think tank, but I suspect this job offered a rewarding trifecta. He could play with all the whacky science stuff. He could develop leaders. He could join the effort pressing for policies that would keep the United States free, safe, and prosperous. It was an offer he could not refuse.

JV's accomplishments in the think tank world have been extraordinary, even for a place where big thinkers are common. Just about the time JV joined us, the air force had begun to deploy the F-35, its newest fighter jet. There was, and remains, a good bit of controversy about the plane. JV has the scientist's zeal to seek quantifiable answers but also the pilot's passion to understand how the plane was really performing in the real world. So he did what no one, not even the air force, had ever done. He interviewed more F-35 pilots than anyone else in the world. He asked all the hard questions that only a fighter pilot would know to ask. He published the findings in pretty much what is the last word about whether the F-35 is a plane worth having. Spoiler alert: It is.[1]

There is no one I know who can make complex scientific and technical issues more accessible, nor anyone else with the vision and practical knowledge to understand how emerging science will affect future life in the everyday world. Having JV around is like having your own science lab down the hall run by your own personal "Doc" Brown (*Back to the Future*, 1986). He could not only build a DeLorean but put wings on it and do a bombing run over Baghdad. John Boyd would be proud.

When I want to understand how science and technology are going to change the future of national security, I talk to folks like JV. If we were all smart, we would listen to what people like JV have to say. The most important thing I learned from my many hours of listening to him is that if you are not thinking about how emergent technologies are going to impact the future of national security, you are already losing.

Below, we will take a look at some of the emerging tech trends that may impact the networking world, but new ones will likely emerge even before this book is printed. So if you want to be a networking world power, odds are, you are a going to need a JV by your side, whispering, "Hey, check this out."

Brave New World

From the beginning of this book, the main point about networking warfare has been that this practice is all about people. Humans drive the competition. Technology is an enabler, like a fighter jet is just an extension of the pilot. That said, new technologies will emerge. Humans will make decisions on how to apply them, and that will reshape global competition. So that means no discussion of networking warfare can end without considering the technologies that may redirect how social networking warfare will happen in the future.

It is always a chancy business making predictions about humans and their technologies. Don't be smug about it.[2] Ask the folks at the Monroe Calculating Machine Company. They unveiled a new computer. They were sure that the Monrobot Mark XI would shock and awe the world. "Up until now, all-purpose computers have required a great amount of space to sit down in and couldn't be readily moved from place to place," the company spokesman pointed out. "Mark XI weighs in at only three hundred and seventy-five pounds." That is not all. "The *real* news about Mark XI is its price. . . . The Monrobot Mark XI will sell for the amazing low price of twenty-four thousand five hundred dollars," he added.

It was 1962. The Monroe Calculating Machine Company had thought it had changed the world.

How much better could things get?

Ask Bill Gates and Steve Jobs.

Of course, *Digital Dominance* might be as wrong about the future of technology as the Monroe Calculating Machine Company (which, by the way, is still in business today as Monroe Systems for Business). Still, there is no option of ignoring what comes next and waiting for it to get here. There are emerging technologies that have already moved from theory and laboratory workbenches to begin impacting how national security competitions will unfold. Just ask JV. He will rattle off a long list that includes everything from blockchain (databases that store encrypted information electronically in digital format) to hypersonic weapons (missiles that can fly up to five times the speed of sound). Get JV started and it is like triggering Bill Nye the Science Guy.

This chapter is going to focus on the emerging technologies that will likely dramatically impact social networking warfare. That raises an interesting question: What kind of technologies would that be?

On the one hand, the answer to that question is easy: All of them. There is no telling what innovation offers the potential to upend the networking world. Blockchain technology is a good example. Most know about its use to produce Bitcoin and other virtual currencies that can be used for payments and cash transfers, twenty-fours a day, anywhere in the world without going through a bank.[3] This, of course, is useful for any network that wants to move money around securely without their activities being monitored and scrutinized. There are, however, many other uses for blockchain that have applications to the networking world. Here are a few.

Digital Identity. Microsoft is pioneering the use of blockchain to develop an app that can issue secure digital identities. "Microsoft's cloud community and developers," one 2018 media report stated, "believe that blockchain represents the perfect way for users to control their digital identities, while also giving users control over who accesses that data."[4] The company continues to develop the technology.[5] This capability could be potentially useful to networks that need a secure means of identifying members of a network.

Data Backup. Blockchain may be used as a tool for secure and reliable data storage. This could well be important for networks that are heavily dependent on data and needing assured access to their data or resilient data storage methods.

Secure Access. Protected entry to physical and virtual space could be important to networks. Blockchain could be used to grant remote access to physical places or digital devices. Super-secure entry might be helpful to either facilitate networking or provide another measure of security and protection for networks.

Furthermore, blockchain is already becoming a tool of great power competition. Adversarial states are looking at the technology to bypass the accountability and transparency of financial services and transaction systems used by liberal states.[6]

Blockchain is a good example of how humans are devilishly clever beings. They are always finding new ways to put new technologies to use. That goes for any technology. That said, there are explicit capabilities in the networking world that could be greatly impacted by next-generation technologies. What kinds of technologies might they be? They will be the ones that address these three capabilities.

Data Accumulation. Data are the nuclear weapon of the twenty-first century. The more data a network has access to, the more potentially powerful the network can be. And, man, are there more data out there. How much is a subject of much debate. I just randomly picked one answer. Statista projects, 74 zettabytes (ZB) of data will be created in 2021. That's up from 59 ZB in 2020 and 41 ZB in 2019. A ZB is a trillion gigabytes.[7] The exact number doesn't matter. What matters is it that there is a great deal of data, and there is going to be a great deal more in the future.

What we do know indisputably is that the increases are growing exponentially. Ninety percent of the world's data have been produced in the last few years. That is a rate of growth that is likely to dramatically increase with the widespread global deployment of 5G telecom systems (fifth-generation technology standards for broadband cellular networks that greatly expand the capacity to move data and more computer processing power closer to the sources of the data) and the new generations of telecom that will follow suit, the continued expansion of cloud services (storage and computing power provided as a service distributed over multiple data centers used for storing data) and the technologies that follow. Wrangling all these data is the Texas-size cattle drive of the twenty-first century. Any technologies that help networks access these data is going to provide a crucial competitive capability.

Computational Power. What makes data valuable is the capability to evaluate information to produce useful knowledge. For many years, we were conditioned to believe that progress in computational power was governed by Moore's law (new generations of silicon chips are produced to double computer speed every eighteen months). Moore's law was discussed in *Wiki at War*.[8] This "law" was more about market practices than science. Regardless, it has run out of physics. There is a limit to how many atoms can fit on a silicon chip. Moore's law is about dead. Furthermore, given the astronomical increases in data, if processing is really bound by Moore's law, processing power will lose a lot of ground in keeping up and we will have way more data than can ever be crunched.

It is not just more capacity to crunch through data that matters. There is a dramatically growing need to process the vast accumulation of information out there more elegantly. A lot of the world's data are unstructured (data that are not organized in a predefined manner or organized in a manner that cannot be processed) or "dirty" (inaccurate, incomplete, or inconsistent data). Technology that can make sense of nonstructured and dirty data is increasingly valuable in a world awash with information.

Any capability and additional capacity to bring more computing power to managing data processing is going to be a boon to anyone with access to data. Having access to world-class data processing for a network is like the difference between giving a quiver of arrows to Robin Hood or Friar Tuck.

Virtual Experience. Since at its core social networking is a human interaction, any technology that enhances human experience in a virtual environment could be significantly impactful in networking warfare.

Virtual environments are, of course, already a thing. Militaries use them for training.[9] Many people experience them through gaming programs. Computer gaming, in fact, plays as dominant a role in modern society as listening to Jack Benny on radio played in the first half of the twentieth century. As a means of entertainment and education, gaming is eclipsing other forms of popular culture technology from movies to music. How big is gaming? Let's ask our friends at Statista. "In 2020," the site reports, "the revenue from the worldwide PC gaming market was estimated at almost 37 billion U.S. dollars, while the mobile gaming market generated an estimated income of over 77 billion U.S. dollars."[10] In contrast, the global cinema box office receipts in 2020 were about $12 billion. Mark Zuckerberg may have made a bad bet trying to prematurely rush his company into becoming the internet's virtual Godzilla, but that doesn't mean virtual technologies won't transform our future.[11] The virtual world is becoming a much bigger part of the real world's reality.

Virtual reality is already playing a role in mediating human social interactions. As one study notes, "Online social games are becoming a significant component in today's social media sites. The social networking sites environment has provided a platform for online games to develop and expand in the virtual medium. Users are now able to play games online, compare scores, and challenge each other among many other things."[12] Some of these interactions are deep and profound. Some users on Twitch (an interactive live streaming service) follow their favorite gamers online for up to ten hours a day. Some online gamers are so consumed by these activities that they require mental health services to adjust to normal day-to-day functioning. Thomas Poell, David B. Nieborg, and Brooke Erin Duffy argue in their recent book, *Platforms and Cultural Production* (2021), that scale and scope of online interactions will reshape business and community activities as well as home life. These social activities can be harnessed for networking activities, applicable in every field of human interaction including national security.

There is already a growing expansion of both malicious and entrepreneurial activity in the gaming space. For several years, for instance, security agencies have been reportedly monitoring gaming sites suspected of being used as tools for terrorist collaboration.[13] According to another report, "a handful of platforms—including the Sandbox, Decentraland, Mirandus, and Axie Infinity—are creating an interactive version of the internet where users can play games, explore virtual worlds, and even do business."[14] These include selling virtual real estate and other assets in the virtual world.

Since games online and offline mimic human learning processes by teaching through experience, gaming can impact the skills, knowledge, and attributes of networkers as well as guiding and impacting the actions they take online and offline. Gaming can be used to teach and modify behavior. Games can help build new skills.[15] They can also allow gamers to plan, test, forecast, and assist in performing real-world tasks.

Any technologies that significantly advance accessing and processing data and putting the knowledge in a virtual presence where humans can interact and manipulate information will change how networking warfare gets done in the future. Networking warriors that don't pay attention to these developments will be fighting the last networking wars not envisioning how to dominate in the next ones.

Shape of Things to Come

Based on our list of the top three priorities, a survey of what's coming in the technology space reveals emergent technologies that quickly jump to the head of the line in offering the potential to reshape the networking world. It is worth taking a closer look at all four of them.

Internet of Things (IoT). The IoT (any physical object embedded with sensors or software that can exchange data with other devices over the internet or other systems) is already a thing. Wearables, like the Apple Watch, for instance, are commonplace. Cars are part of the IoT universe. An average car today has more connectivity than an Apollo spacecraft. Many US homes have smart devices that do everything from turn off the lights to sharing cookie recipes. How big is the IoT? Another one of those "who knows?" questions. One estimate was that the IoT business sector was almost $309 billion in

2020, expected to rise closer to about \$2 trillion by 2028.[16] Guessing that estimate is too conservative, but we'll see. The point is, the trend line shows we are going to see a great deal more IoT. There are already more IoT devices than humans on the planet. The IoT is not just about having a lot more devices; it is transformative. Here is how.

For starters, the IoT is changing how everything connects. "Nowadays, the main communication form on the Internet is human-human," wrote researchers Lu Tan and Neng Wangin (2010), "but it is foreseeable . . . that any object will have a unique way of identification and can be addressed so that every object can be connected. . . . The communicative forms will expand from human-human to human-human, human-thing and thing-thing."[17] This is not to argue that human networking won't remain the core of social networking warfare (I'm placing big bets that it does). What is particularly attractive about the potential of the IoT is the greatly expanded possibility to collect and move data. It is little wonder that the emergence of the IoT happens to coincide with the explosion of the volume of global data being produced. It is little wonder as well that when IoT deployment is married with 5G and the communications technologies that will follow delivering future developments in data storage, management, and retrieval, data availability is going to go off the charts.

The IoT will bring a host of perils and possibilities. For starters, networks dependent on data will be much more vulnerable to denial-of-service attacks (a threat discussed earlier). The IoT greatly proliferates the number of potential zombies to generate botnets. The IoT also controls a growing spectrum of devices that manage everyday life from the cars we drive to the thermostats that can shut off heating in our homes in mid-winter. Networks that are particularly vulnerable to disruption from denial-of-service attacks will be at risk. New countermeasures will have to evolve to ensure network resilience.

Another growing area of concern will be issues relating to privacy and civil liberties. The last chapter addressed rising fears of big tech having access to personal data. That challenge is magnified many times by the IoT since big tech companies are behind many of these devices and have access to the data they collect. The behavior of states will also be a cause for concern. China is a case study. The regime is already harnessing the IoT to enhance the capabilities of authoritarian surveillance states. One issue attracting attention is the deployment of "smart cities" technology. China is a world leader in marketing technology systems to help manage utilities, logistics, traffic, and security, all infrastructure that can also be manipulated to control urban populations.[18]

On the positive side, the IoT will also likely create a plethora of new goods and services.[19] This will, in turn, expand the consumer and civil society space online, opening new opportunities for networking activities. As the IoT proliferates, it will expand where and when networking can occur. For instance, a few years ago it would have been unimaginable that companies could hold board meetings in a car on the way to an airport, rather than taking cars to go to an airport to attend a board meeting somewhere else. Today, virtual meetings on the move are commonplace. As the geographic and temporal space for networking expands, so too will networking opportunities.

The IoT will without question deliver more data to networks in more places. Any network that expects to dominate social networking warfare in the future ought to take that into account. To not would be like an admiral not taking aircraft carriers seriously after Pearl Harbor.

Web 3.0. At least one commentator with a good deal of experience in public policy and public affairs, Scott Zipperle, argues that a combination of technologies represented as Web 3.0 will break big tech's monopoly of digital dominance on the internet.

There is no universal definition of Web 3.0; the term is shorthand for what will emerge as the next suite of user interface tools on the Internet.[20] In our exchange, Zipperle pointed to three capabilities that could be transformational—data ownership, decentralization, and algorithmic transparency.

Data ownership would grant users ownership of their data by default, presumably by using blockchain or some similar tool to safeguard information. This would create a level playing field in the digital space that would allow users to become brokers of their own information.

Decentralization could come from new technologies that disrupt the traditional systems of distributing and storing information. Decentralized storage and data distribution would democratize the information management space.

Algorithmic Transparency. Algorithms (the rules computers use to process information and problem solve) are the secret sauce of digital space. The lack of transparency creates a double challenge. On the one hand, algorithmic bias can be used to drive outcomes rather than let the data speak for themselves. On the other hand, the lack of transparency can be used as an excuse for others to step in and declare their own outcomes as more equitable and just. For instance, "claims that machine learning algorithms disadvantage women and minorities are commonplace today," points out Stephen Baker, an expert on cyber and privacy policy. "So much so that even centrist policymakers agree on the need to remedy that bias. It turns out, though, that the debate over algorithmic bias has been framed so that the only possible remedy is widespread imposition of quotas on algorithms and the job and benefit decisions they make."[21] That remedy, however, just substitutes one bias for another. There is only one way out of this digital death spiral. Creating open-source systems or means to audit, inspect, and oversee how algorithms process data would allow users to scan for bias and other stratagems that are used to manipulate and influence how data are processed and presented.

Artificial Intelligence (AI).[22] AI (computer systems able to perform tasks that normally require human intelligence) is to data like the Cookie Monster is to cookies. AI offers a great leap forward in the capacity of computers to create knowledge from data and do something with it. Rather than supplant humans, AI will likely dramatically enhance human activity, and network warriors could be one of the greatest benefactors.

The ability of computers to evaluate information, make choices, and act on decisions has made remarkable progress over the past decade. One area discussed previously is the development of data mining. Data mining looks for patterns in data. Advertising agencies have used it for decades to determine which campaigns have the greatest draw and to identify specific target audiences. Like many other techniques originally carried out with pencil and paper, data mining has become faster and easier with the use of computers. Coupled with technologies that allow better gathering of raw data—everything from laser scanners at supermarket checkout counters to unmanned aerial vehicles on the battlefield—the volume of data available to decision-makers has increased dramatically.

What networkers really need to pay attention to in the near term is a particular method of AI called machine learning. Machine learning is technology created by developing processes that mimic human brain functioning, patterned on how brain cells work in a neural network. This approach to computing is "data driven," providing inputs that became the basis for establishing cause-and-effect relationships in a similar manner to how human brains create knowledge and make judgments.

Although machine learning may not be the technology that delivers the most advanced forms of AI in the future, its impact on contemporary developments in the field are unquestioned. Machine learning will, without a doubt, be part of a family of technologies that delivers a new generation of computer services. For example, pairing computers that can make better decisions with sensors, like IoT devices, that can collect more and better information will produce new capabilities, synchronizing the benefits of advancements in both technology fields. In short, machine learning technologies will likely be ones impacting networkers the most in the near term.

In the next five years, machine learning–enabled technologies that can deliver reliable, scalable, cost-effective capabilities are going to wash over the marketplace like a tsunami in many fields in the private and government sectors. Today, machine learning capabilities are already ubiquitous in many widely deployed technologies, including speech and facial recognition. When the final numbers were tallied, one market research report estimated that total business worldwide, "including software, hardware, and services, are expected to total $156.5 billion in 2020."[23] This would represent more than a 12 percent increase from the previous year—remarkable growth, given the drag on the global economy from the COVID-19 pandemic.

Where machine learning technologies will have the greatest influence on networking warfare will likely be where these technologies complement rather than supplant human work, blending the work of machines and humans. In networking warfare, humans and computers will be working in a collaborative environment. This will be the dominant structure in the future structure of networks. The reason is that no other systems will match their capacity to process the large volumes of unstructured data that are going to be available and from that mass of information produce knowledge and dominant action.

The relationships between AI and human networkers will be multifaceted. Thomas Malone, director of the MIT Center for Collective Intelligence, postulates that work structures will be dominated by three types of human-machine collaboration.[24] (1) Machines serve as assistants. This is already happening. One example of this relationship can be found in computer-assisted surgery. (2) Machines function as peer partners with humans. One team of researchers postulates that "by combining the two concepts of assistant systems and cognitive architectures, we can create a system which is capable of seamless human-machine interaction and integration, like a peer to [its] user instead of a servant or a simple assistant." (3) Machines function as managers, directors, and supervisors. Machine learning technologies are already employed in reviewing and making recommendations on human performance, such as reviewing prescriptions issued to pharmacists.

The areas where AI will influence networks are manifold. For example, AI is already being employed in the defense of networks, providing new layers of protection in cybersecurity. Machine learning technologies are already being used to help identify cyber threats and direct countermeasures. All these methods are heavily dependent on automated processes that involve little or no human intervention. The proliferation of machine learning–enabled technologies in this field is likely to continue to expand rapidly. The trend of networkers working in a distributed fashion remotely on diverse digital devices and systems will likely even further accelerate the demand for "smarter," scalable, and proliferated AI-empowered cybersecurity. Since AI technologies will also enable cyberattackers, integrating AI into cyber systems will be essential to ensure reliable, robust, and resilient computational services and communications.[25]

AI will also greatly power networking action, both by generating knowledge from data and facilitating collaboration to make and implement decisions. The efforts will be enhanced by the intermingling of the IoT and AI. Machine learning is already widely applied to software for IoT devices and services to make them smarter, more secure, and more productive. In turn, since machine learning benefits from large volumes of data to operate successfully and improve performance, and networks of IoT sensors and devices provide an enormous amount of information, the synergy of the two create added values to business, industrial, consumer, and government applications. Whereas the benefits of machine learning and the IoT are already established in consumer products, the potential impact on network activities could be even more far reaching.

AI, however, won't serve to solve every problem in networking competition. There is a simple reason for that. For AI to work best, it needs a lot of data. Unfortunately, not every problem to be solved is data rich.

Machine learning technologies are more effective if they have a lot of data where they can learn well-established patterns in bound environments. A good example is traffic systems, where computers could learn from past commuter behavior to manage future traffic flows. However, national security competitions tend to be highly complex and chaotic, often involving very big activities with very limited datasets. The 9/11 attacks, for example, are a dataset of one. Everything the White House knew at the height of the Cuban Missile Crisis could have fit in a binder. Sometimes, the biggest national security challenges are the most data poor. Then all the AI in the world won't give you all the answers you need.

There is another reason AI might come up short for national security networkers. Human choices rather than the proliferation of technology will likely drive how these technologies are employed. Like all new technologies, their deployment introduces new possibilities and perils. Here is a short list.

Privacy. As with other advanced technologies, AI raises potential concerns about abuse of privacy and other civil liberties. One example is public anxiety over the use of AI-based facial recognition technology by police departments. Several leading tech vendors, including Microsoft, IBM, and Amazon, announced that they would limit sales of facial recognition software to law enforcement agencies. Another concern is use of machine learning tools to create "deepfake" misinformation efforts (a threat cited earlier).

Laws. Regulatory structures will also significantly constrain or speed up the deployment of new AI technologies. Potential new regulatory structures for machine learning–enabled technologies could cross over into labor, environmental, safety, security, privacy, and many other regulatory frameworks that may become integrated into the rules governing AI technologies. There is a no comprehensive regulatory framework for AI-enabled technologies. Regulatory frameworks that govern the internet were developed under extant laws governing commercial telecommunications networks. Unlike the internet, AI is a new generation of hybrid capabilities that cannot be treated like telecommunications technologies. Managing them will require something new. And whatever new looks like, it has to enable, not undermine, the development of new AI capabilities. Misapplication of regulatory structures could undermine innovation, limit competitiveness, and inhibit growth.

Cooperation. Levels of public-private cooperation and collaboration in AI development will also be key. As noted previously, no aspect of public-private cooperation on emerging technologies has greater import than how the United States will engage in great power competition, particularly with China. China is a powerful competitor in this sphere not only because of the size of its economy and its capacity to exploit technology but also because of the Chinese practice of military-civil fusion (mentioned previously) that allows the Chinese Communist Party to seamlessly integrate all elements of national power, including industries and companies, to press the developments of its AI capabilities. China has identified AI as a key strategic development. The United States and other liberal states won't be able to keep pace without efficacious cooperation between the public and private sectors.

In summary, AI is the answer to the future firestorm of data coming at network warriors. The future networks that dominate will be AI enhanced, or they won't be dominating for very long.

Quantum Computing. Harnessing the power of the properties of quantum states offers a mind-boggling potential computational power beyond what can be accomplished with binary computational computing. Here is what matters to networks. Quantum computing is the third leg of the information superpower trifecta. If developments like 5G, cloud services, and the IoT can deliver unprecedented amounts of data and AI can process that data, quantum computers hold out the possibility of conducting that processing at unprecedented volume with astonishing speed. In the world of networking warfare, the difference would be as stark as a high school track star in a scratch race with Usain Bolt.

Without getting all physics professor about how quantum states work, the best explanation I can offer is from another of my think tank big brain buddies, Klon Kitchen. Klon explains, "Quantum computers use the unique properties of quantum mechanics to enable computers that are exponentially more powerful at certain tasks than any supercomputer ever built. Conventional computers, at their most basic level, store and use information as individual bits, which encode information as either a '0' or a '1.' Computer programs do what they do by manipulating millions of bits in different patterns to accomplish different tasks."[26] In the computer world, a functioning quantum computer would be to the computer on your desktop as the most powerful supercomputer in the world today is to a paper and pencil.

Unlike the other technologies covered so far, you can't go to the Amazon store and buy this technology (though you could buy a book on quantum computing like Robert S. Sutor's *Dancing with Qubits: How Quantum Computing Works and How It Can Change the World* [2019]). In fact, the jury is still out on whether all the practical obstacles to building quantum computing at scale is achievable.[27] There are, however, folks working on this. For sure, you know who thinks quantum computing is not only possible, but an absolute game changer? China. China has undertaken an enormous research and development in this space. In partnership with its capabilities in 5G and AI, the regime expects that together these capabilities will make China an unassailable information superpower.[28]

The potential benefits are still theoretical, but the theories are compelling. In addition to unprecedented computational power, quantum computers could be, for instance, foundational to a more secure internet, as one research team at Princeton concludes. They are "working to transmit quantum information from photons to electron spins, where further fine-tuning can prolong the quantum state by keeping electron spins in the proper orientation. Quantum entanglement ensures that this new kind of internet is secure against hackers. Any attempt to eavesdrop on the transmission will perturb its state. By comparing the transmitted photon to its entangled twin, the receiver can tell if an eavesdropper has disrupted the transmission. As long as the laws of physics are correct, our channel is secure."[29] In addition, there could be other features of quantum architectures that offer enhanced computational powers enhancing any aspect of analysis that requires massive numbers of calculations and generating probabilities.

Applications to national security competition are extensive from predicting weather on a battlefield to the resilience of cities under cyberattacks.

Any of this knowledge might be massively useful to networking warfare, speeding the assessment of situations, decision-making, and implementing actions. Quantum computing may well one day be the ultimate expression of Boyd's observe, orient, decide, act loop with decision cycles moving close to the speed of light. Captain Kirk (*Star Trek*, 1966) would be jealous.

In the end, money, brains, and physics will decide who develops the unprecedented potential of quantum computing. The biggest challenge is big, powerful states putting their fisted hand on the scales. Liberal states need to worry about losing this competition to China.[30] Forget the race to build the atom bomb or to get to the moon; this fight could be more consequential to the future of freedom than any of them.

As with many aspects of tech competition, much of the capabilities for moving quantum computing forward will rely on the capacity of the private sector. As noted previously, liberal states compete differently than authoritarian regimes. Furthermore, mimicking the practices of authoritarian regimes is likely to make liberal states less and not more competitive. So, that raises the obvious question: What can liberal states do to ensure they are in the best position to compete? Here are three answers.

Understand the Competition. Governments can serve an important role providing situational awareness on the state of competition and research and development in the field. This should be a multifaceted effort including the intelligence, scientific, and technical communities.

Collaborate, Not Dictate. Centralized research and development planning will likely limit innovation. Just throwing money at research doesn't necessarily guarantee progress either. Furthermore, the applications for quantum computing will have as much if not more application in the commercial realm as in the national security space. What governments ought to do is proactively push for the trifecta of cooperation between the resources of government science and technology, academia, and the private sector.

Protect Intellectual Property. Much of the innovation in this space will represent cutting-edge science and engineering. Protecting that work is an appropriate and important role for nations. "Bolstering U.S. protections for intellectual property (IP) rights will be a fundamental part of becoming the preferred environment for quantum research and innovation," argues Klon Kitchen.[31] This includes harnessing all the relevant instruments of government from granting patent rights to investigating and prosecuting offenders. What is also important is for states to be an active international actor pressing other states to meet their international treaty obligations and bilateral agreements that protect intellectual property.

What remains to be seen if this technology does mature is when and how it is proliferated. We could well see capabilities distributed much more widely and rapidly than supercomputers. On the other hand, it is not likely in the near term that everyday consumers will have ready access like personal computers and other digital devices. That said, it is probable that one of the first places that quantum computing power is directed (outside basic scientific research) will be in the field of national security. Therefore, within the near future, quantum computing could begin to significantly impact national security networking warfare.

Quantum computing may be the least mature of emerging technologies, but its potential is the most far reaching. Serious networking warriors who do not prepare for a world with quantum computing do so at their own peril.

Virtual Reality Technologies. This category is not a single kind of technology but a family of capabilities that create a simulated physical environment (real or imagined). It may include software, sensors, motion detectors, gyroscopes, and visualization and projection technologies. How much appetite is there for these technologies? Well, according to Statista, it's a big number—projected at over $300 billion in 2021.[32] For context, the entire US automobile industry is not worth one-third of that. Future growth projections for virtual technology are exponential. Any carmaker would kill to have growth projections like this.

Although virtual reality technologies are often associated with online gaming, immersive technologies are already widely used in fields from medical science to military training, engineering, and education. They are mostly of interest to networking warriors for their potential to shape virtual reality environments which will facilitate an understanding of challenges and solutions to national security problems that can be studied and rapidly tested. This family of capabilities could well prove to be the capstone of an emerging assembly of technologies that will reshape the future of networking. Other technologies will produce, analyze, and deliver unprecedented knowledge based on an unprecedented amount of data. What virtual technologies potentially provides is the ability to put this information in a format more accessible to humans. Ultimately, since networking warfare is primarily a human-to-human activity, technologies like virtual reality augmentation could have a powerful effect.

What might new generations of virtual reality technology deliver? Without question, especially as 5G and other communications technologies move more data and push processing power closer to the user, virtual augmentation will become accessible to a wider variety of digital devices including traditional digital platforms as well as IoT devices.

Technologies are also likely to provide increasingly rich and realistic immersive environments. The speed of creating these environments will also increase significantly; that may prove to be the most crucial innovation of all. One survey of technology predicts, "Simultaneous Localization and Mapping is the biggest focus when it comes to development at the minute: this technology is able to instantly translate data from the real world (by using sensors) into the virtual one and vice-versa."[33] Additional advances are also anticipated in 3D virtualization presentations. Much of this work is already being pioneered in medical sciences.[34] The cumulative impact of these technologies will allow for rapidly establishing immersive environments tailored to the needs of networks, adapting in near real time to their requirements.

This is not to say there are no significant challenges to duplicating the real world in virtual space. One is latency, the time delay between the cause and the effect of physical change being observed translated into the virtual world. "The gaming industry can, and is very good at, developing latency fixes for all gaming elements in their 'simulated' worlds," JV told me. "But simulating real-world operations, like linking hundreds of advanced fighter simulators together, is proving a real challenge." In short, for big, complex activities, simulations may not be able to duplicate picture-perfect reality in a manner that is required for some operations, analysis, and decision-making.

Much of the most exciting developments in this space are likely to happen in the commercial sector. Facebook, now called Meta, has bet its future on creating the "metaverse," a fully collaborate virtual environment that they anticipate will become as common a platform for social interaction as the internet.[35] It now looks like the first efforts of the company in realizing their vision of re-creating *The Matrix* (1999) flopped; this effort undoubtedly reflects the face of social networking to come. National security networkers will be operating in this environment. That is a given. They, even government networks, will likely largely be leveraging private sector goods and services. The plain fact is that national security networkers should be giving as much attention to developments in this space as they are to other emerging technologies.

Speculation

Many of the technologies described above are not only rapidly maturing, but they are also being fused to deliver new capabilities. The US Air Force, for example, is going to spend about $1 billion pairing AI with gaming technologies to predict and manage air flows in combat operations.[36] The future is happening.

If all the technologies described here were available now, how would that impact social networking warfare? To think about that for a bit, let's go back to the start of the book and recall the last bad days in Afghanistan when our ad hoc network, along with many others, scurried to help innocent people escape the Taliban. How might that have looked different?

For starters, instead of trolling through Outlook to search "who did I know who might be able to help," I might have had access to data and computer processing that could have done with an instant query of known relationships and using AI-enabled software in real time to map out a suggested network for the tasks that needed to be accomplished. That capability alone would have gotten our efforts twenty-four to forty-eight hours ahead of the problems we were trying to solve.

Advanced virtual reality simulations could have allowed for visualizing the entire sequence of activities from getting folks to the airport, through security, and onto a plane. These simulations would be available for use and updated constantly even as arrangements were being made to contact people for evacuation. We would be running different scenarios in real time on the best courses of action. The simulations would update as the world of data was swept for new geospatial information and other important insights. Data from IoT devices would be streaming in reporting on everything from when people ate last to how much cash they have in their pocket. We could track in real time the progress of evacuees on the ground. We could move money virtually instantaneously to pay for the goods and services needed to clear the way. We could, the whole time, have a virtual common picture of situational awareness for the entire network. Instead of moving at the speed of Outlook, we could have been moving closer to the speed of thought.

Would everything have gone infinitely better? Maybe not. Others would no doubt have had similar capabilities. As we were fighting to get people out, they might be meddling to shut things down. No doubt, we would be back to the timeless game of competition—a contest of wills, a struggle of action and counteraction, measures and countermeasures. What would no doubt be true in this future place is that if we were not maximizing these technologies, we would be losing. For the simple reason that others will use them.

Future of the Future
How networkers ought to think about the future is a subject given more attention in the conclusion of *Digital Dominance*. But what is worth reminding in this space is that networkers that want to keep winning and dominating must not only think about how to win today in this environment but ponder

over how to win in the future as the world and as technology changes. That is why you will need a JV-like capability if you want to keep winning.

Who knows if these technologies will ultimately be the ones to reshape the future of networking online and offline. What there should be no doubt about, however, is that future networking will likely be shaped by the cumulative impact of a family of technologies rather than driven by the development of a single product, good, or service. What makes sense for futurists is to think holistically about how the constellation of emerging technologies will impact the future of networking warfare online and offline and, most important, what they or their competitors might do with the new tools in their hands.

EPILOGUE
THINKING THE FUTURE

Rest in peace, Jim McGann.

While I was writing the conclusion to *Digital Dominance*, my network of think tank folks started echoing the sad news. James G. McGann, the long-time director of the Think Tanks and Civil Societies Program at the University of Pennsylvania (mentioned in chapter 12) died unexpectedly.

I think Jim would have appreciated my using this space to say goodbye to him and remember his leadership and scholarship. Dr. McGann's work researching the think tank community worldwide, sharing ideas and best practices, had global impact. As *Digital Dominance* argues, the think tank and nongovernmental organization community (chapter 12) have an important role to play in keeping the world safe, free, and prosperous. His work made the community better.

Reflecting on Jim's life and contributions is an appropriate starting point to contemplate on what to do with *Digital Dominance*. This book is about the things Jim thought were important—hard thinking about where the world is and where the world needs to go, and the power of powerful ideas welded with concerted action needed to make that happen.

On the one hand, it is a melancholy bookend to the writing this book started with, recounting the tragic day America abandoned Afghanistan. We started with a loss. We end with loss. On the other hand, it is another affirmation of why this book was written. We need strength for a fight we can't afford to lose. If the free world wants to be free, the free world is going to have to fight for freedom.

The battlefield that *Digital Dominance* focuses on is the world of social networking warfare both online and offline. These are first and foremost combating human networks, arenas of competition as relevant to our future as Agincourt (October 25, 1415) was to the reign of Henry V. The great power of connecting in the modern age comes from the action that flows when the real and virtual worlds are connected and endowed with the power to influence events. In this new world of competition, there are better ways to win. *Digital Dominance* laid them out step by step—how to build and operationalize a network to deliver dominating action. This book also covered how to win by making your enemy lose—disrupting, defeating, or destroying adversarial networks. *Digital Dominance* also described the real and digital places where the fight for dominance is taking place among state actors, nonstate actors, and the private sector, as well as the role technology may play in shaping the competition.

Things to Think About

How long this handbook for dominating networking warfare remains relevant remains to be seen. After all, the world of national security is a world of competition. In the end, it is the competitors that make and remake the world where they compete. They are always trying to rewrite the rules to make sure their side survives and thrives. The advice offered here may not survive the test of the future. So *Digital Dominance* ends with adding some intellectual insurance that just might help keep you ahead of the competition.

Digital Dominance concludes with things to think about concerning what the future may bring. Here is a short list. These are topics that are not primarily about networking, but they could well determine how well networkers network in the future. There are four of them.

Energy and Environment

Worry about the debate over climate change policy. Some would argue climate damage from human activity is an existential threat to humanity and our host planet. They argue for directed, transformative action that supplants all other concerns. Others argue it is all nonsense. Still others contend this is the only blue dot we have. We ought to take sensible steps to preserve the environment and the planet's resources, but we ought to find efficacious ways to do that that do not undermine human freedom, security, and prosperity. How this debate is resolved is going to have a huge impact on networking warfare because in all our lifetimes, there won't be much networking (beyond hunting and gathering) without electricity.

As energy and environmental issues gain greater prominence in public policy, they are more likely to play a significant role in shaping the use and availability of current technical systems and the adoption of new technologies. In the near term, digital devices are going to be mostly powered by electricity. That seems as indisputable a projection as death and taxes. By some estimates, computer technologies could consume over 50 percent of global electricity demands by 2030.[1] To be honest, we really don't know how much electricity we are going to need. For instance, one media source noted, "If every search on Google used AI similar to ChatGPT, it might burn through as much electricity annually as the country of Ireland. Why? Adding generative AI to Google Search increases its energy use more than tenfold, according to a new analysis."[2] How nations elect to address climate mitigation could dramatically affect the cost and availability of electricity.[3]

Emerging technologies could also affect future energy supplies. For example, efficiently meeting electricity demands is heavily dependent on forecasting. Forecasting technologies are critical to "capacity planning, scheduling, and the operation of power systems."[4] In short, it may take more technology and more energy to produce better environmental outcomes. Bad policy choices could well constrain not only energy supplies but also emergent technical solutions that could provide better ways of doing harder things.

Worry about the climate debate, because the arguments have become overly politized, with agendas mattering more than outcomes.[5] Rather than thinking about environmental issues as a global crusade or an existential crisis that trumps every concern and national interest, climate and related challenges can be dealt with most effectively by treating them like other domestic, national security, and foreign policy problems, pairing realistic assessments of the challenge with suitable, feasible, and acceptable solutions.

In liberal states, the Left and the Right agree that technology can be a powerful tool for building a better world. Why wouldn't we all want a new generation of green technology that delivers high productivity, greater efficiency, reliable and abundant energy, and better environmental outcomes? A centrally directed governmental program will not deliver on that agenda. In fact, government intervention often stymies innovation and misallocates public and private dollars, meaning economically and environmentally promising technologies, policy innovations, and creative solutions may get left behind.

Technological change and innovation on a grand scale should not be deterministic. How we have described the role of technology in networks holds true for energy and environmental technologies as well. Free nations disadvantage themselves when they compromise on the benefits provided by a free society. Empowering politicians to pick winners and losers in economic competition is as likely to lead to widespread corruption, inefficiencies, and corporatism as it is to produce optimal economic and environmental benefits.

That is not to say that there is no role for government, but that role should be limited to legitimate matters of governance—protecting the rule of law and human liberty, not supplanting market choices.

Education

The best networkers are humans. The best human networkers have better skills, knowledge, and attributes than the humans they are networking against. Skills, knowledge, and attributes are the product of an education. Educating a networker is a confluence of three activities: (1) teaching a skill, knowledge, or attribute, (2) creating opportunities through practical experience to apply skill, knowledge, or attributes in the real world and hone abilities, and (3) having an accreditation or validation that a networker is proficient in the necessary skills, knowledge, and attributes—and they are ready to both be and lead networking warriors. The current formal educational structures in liberal nations are mostly marginal at doing this. I mean, we are not terrible; we delivered the likes of Microsoft, Apple, and Amazon. Still, we are marginal—and it looks like we are getting worse. What concerns me most is, again, the problem of politics.

Educational systems are increasingly being recruited as spaces for cultural, political, and ideological warfare rather than institutions to teach individuals the fundamentals for operationalizing their capacity to act as humans. Educational places that are dominated by politics are going to deliver political outcomes, not educational ones. Furthermore, simply renaming political goals as educational objectives is not educating people—it is indoctrinating them. The degradation of the educational process is undermining all three components of the educational experience: diluting educational outcomes, disadvantaging career opportunities, and corrupting accreditation.

For liberal states to win, they are going to have to out-educate adversarial states. Like with most aspects of competition, mimicking how authoritarians educate, by centralized command direction, won't likely produce better outcomes. Rather, if the liberal world wants to come out on top, it is going to need depoliticized, versatile, dynamic education models that deliver credible teaching, experience, and validation of the linear and nonlinear talents that make for world-class dominating networkers.

In the pages of *Digital Dominance*, I have introduced exemplars, some of the best-of-the-best networkers. They are at first glance a diverse and rowdy lot, an eclectic collection of men and women, seemingly only united in their capacity to conduct esoteric actions from killing terrorists to conquering talk radio. In their biographies, however, there are some common elements that are worth noting. They each in their own way have a combination of solid academic and real-world experience, sharpened by operational practice, and validated by their peers as people of distinction. They ought to give us confidence that building a professional development model that emulates their success could deliver the real leaders we need.

Human Flourishing

Foundational to becoming an educated person is becoming a person of character and ability. For starters, demography matters. After all, first you must have people to make sure you have the best people. Population shifts can significantly affect productivity and the capacity of nations to compete. For instance, what is often called the "East Asian miracle," the explosion of economic growth between 1965 and 1990, occurred, at least in part, because of the region's demographic transition with the working-age population growing at a much faster rate than the dependent population.[6] Migration patterns have also greatly affected productivity and caused shifts in workforce composition and activities.[7] The population of the liberal world (with the exception of places like India) is declining. That's a problem because in the end, nations with more humans can do more human networking. I said it before, but I will say it again. The old Napoleonic military adage is true: "Quantity has a quality all its own." Winners are going to need people. Winning nations are going to have to be better at adding to their population through both immigration and domestic population growth.[8]

Quality also matters. There is no substitute for having a healthy, prosperous, hardworking, and virtuous population. They will not only be the best networkers; they will also be the best everything. Again, I fear contemporary politics in the liberal world is making meeting this challenge harder, not easier. Inserting political agendas into the middle of family building and child-rearing isn't helping. There is strong fact-based case to be made for what is called the success cycle: education, work, and family formation.[9] This structure was once considered noncontroversial. Now people have fistfights over it. Folks may not like this model. If they want to promote others, fine, but they ought to be able to demonstrate better outcomes. People need to start making better choices—ones based on outcomes and not politics. No matter how societies decide to raise future generations, if they don't start doing a better job, those generations won't get to shape their future. Others will do it for them. They won't like what they get.

Mental health matters as well. The more the scientific and medical communities dig into the impact of an assault online activity on the brain, the more warning signs emerge. As one study noted, "As social media started gaining popularity in the mid-2000s, the mental health of adolescents and young adults in the United States began to worsen."[10] In addition, we ought to be wary of destructive new behaviors that could emerge. One innovator recently suggested that the virtual world will allow for the development of "digital babies" as a satisfying alternative for young adults to raising actual children. My colleague Scott Zipperle argues, "It would be gravely unwise of us to shrug off something like raising a digital baby in the metaverse as just a fad that will come and go or to dismiss it as a distant problem for another generation to deal with." He adds, "What it says about us is deeply unsettling: that we are a culture of lifestyle as opposed to a culture of life itself."[11] Some folks may disagree with Scott. That's a worthy debate. Arguably, many of our modern health challenges stem more from our choices in constructing contemporary culture and modern society, rather than just a problem caused by the internet. What has to be acknowledged, however, is that we have to start thinking about mental health at the front end of forging our brave new world, not as an aftermath.

Future work matters even more. We are simply going to have to do way better at building a workforce that can dominate in the networking world, skilled at linear and nonlinear tasks and empowered to the maximum. This is an issue provocatively addressed in Max Smeets's *No Shortcuts* (2022). He looks specifically at the capabilities of military cyber-forces and issues including personnel, actions and analytical tools, infrastructure, and organizations. The discussion of these challenges, however, are widely applicable to all kinds of network warfare.[12]

Future Think

There is one thing I know for sure about all the challenges I have raised here: If they get solved, it will be in the future.

The future will be here before we know it. We had better be prepared for the occasion. I don't think most liberal states think about the future very well. That is weird. We are producing more knowledge all the time. You'd think we would be better at using it for thinking. Most of us are not. That's a problem for networking warfare because in the end, winning at networking warfare is about outsmarting your enemy.[13] How did we get in this mess?

Today, billions of users are on the World Wide Web. Modern researchers have access to vast digital libraries and databases as well as powerful search and computational programs. New means of manipulating data, such as informatics (the science of information processing); data mining (extracting and analyzing data to identify patterns and relationships); computer modeling and simulations; and open-source intelligence (acquiring and analyzing information from publicly available sources to produce action-able intelligence) are delivering revolutionary instruments of knowledge discovery. Nevertheless, while the means of knowledge discovery have be-come more sophisticated, the process of public policymaking (especially for national security) has become increasingly intuitive, and that's right the old bugbear—politicized. People who say, for example, "Trust the science" are really saying, "Trust my political views which I call science." In Washington, for instance, talking points, gut feeling, partisan preferences, and ideologi-cal fervor crowd out cutting-edge, multidisciplinary analysis.

In part, the rise of political platforms over analysis in public policy debates is part of a profound transformation in our understandings of the represen-tation of truth and facts driven by postmodern philosophy and literary criti-cism. "These have led scholars to value 'smart' and 'interesting' work over the 'sound' and 'rigorous' studies that were most praised in earlier decades," suggests sociologist Michèle Lamont.

What was once a new fashion in some academic circles has become "com-mon sense" in debate in the public sphere and weaponized by political fac-tions—inventing their own public policy and inventing a reality to match it. This has led to manufacturing and decrying "misinformation," "fake news," and "fact-checking" as defined by self-defined standards that match what they think, not what we legitimately know through a process of critical think-ing and analysis.

If the free world lets the people of politics do their thinking for them, there is trouble ahead. Futures thinking is particularly susceptible to manipulated judgment. It is way too easy to just pick the future you believe in, like tapping your heels three times and saying, "There is no place like home."

What we need instead, not just for thinking about networking warfare but pretty much everything, is a more rigorous, depoliticized multidisciplinary approach recognizing that there is no assured single path to knowledge. We need to tap the full linear and nonlinear skill set. This approach argues for testing cause-and-effect relationships through several means. The concept of multidisciplinary studies is not new, but this approach can be particularly fruitful now. The information age provides an unprecedented capacity to tackle tough problems in different ways with more rigor.

On their own, any one analytical method might still be an imperfect means for evaluating available data even with the power of quantum computers. In the real world, real problems are plagued by no data or "dirty data"—a conglomeration of incomplete, undependable, ambiguous evidence that de-fies easy analysis. Combining various ways of looking at the same problem together, however, provides policy analysts with a richer and more nuanced view of how to interpret the facts before us. In the end, the answers provided might still be unclear or contradictory. Decision-makers might still have to make intuitive judgments, but they would at least be able to make them with the confidence that their assumptions, predispositions, and preconceptions had been rigorously put to the test.

Some of the different methods and structures for analysis have been introduced in *Digital Dominance*. There are also many other diverse and insightful ways to think about the future than just a linear extension of the present. These methods include everything from scenario-based planning (looking at alternative futures) to horizon scanning (assessing anomalous futures projections) to the Delphi method (structuring the analysis of experts).[14] It is worth using the whole analytical tool kit to figure out what is coming rather than just picking the future you want.

All that said, here is a prediction about the future: The better future thinkers will more profoundly shape the future.

Networking Warfare Agenda

The epilogue of *Digital Dominance* adds more to think about on top of the challenges of winning in a socially networked world. Here, we raise the final question the book will address. The work has covered how to win. In conclusion, I also want to answer the question, how does the side I want to win win?

I confess a prejudice (held for the end). My hope is that the liberal world comes out on top in the social networking war. Here is why: I expect liberal nations will do a better job preserving human freedom, protecting life, being better stewards of the planet, and advancing prosperity. That is a place where I would prefer to live and the world I would most like future generations to inherit.

Here is the problem: Everyone on both sides will read *Digital Dominance*. Well, maybe not everyone on both sides will read the book or even hear about it, but there will be many on both sides practicing the ideas in these pages because they will all desperately be trying to win.

Knowing how to win, however, is way different than winning. The difference is that winners do it—and do it better. So here is the winning agenda for the free world. Many of these proposals have already been introduced in *Digital Dominance*, but to finish, let's put them on a to-do list.

(1) Have productive societies. Networking can be done from anywhere. Networking warfare works better if the networkers are working off the foundation of a strong, resilient, secure society; that includes an economy,

the technology, a population, an infrastructure, and a national security capability equal to the nation's interests. Start with policies that keep the nation free, safe, and prosperous.

(2) Have public-private partnerships. The greatest capability that free nations bring to the table of competition is a vibrant, innovative, dynamic private sector. To thrive and survive, the public and private sectors need to work in trust and confidence, under a system that respects the rule of law, the principles of free enterprise, and the transparency and accountability demanded by democratic governance. This is a relationship that must be based not on politics but on protecting vital national interests. Work in trust and confidence and never drop your virtue at the door.

(3) Empower the arsenal of ideas. The swing voters in the fight for global mastery are the instruments of civil society, including think tanks and other nongovernmental organizations. Authoritarian states can never tolerate a free and open civil society. That is a weapon of networking warfare that can only fully be taken advantage of by free nations. Never leave that decisive advantage on the table. Winners freely think.

(4) Build networking tribes. Networking is a human business (how many times did I write that? Might make for a good drinking game). A generation of networking warriors is one of the best guarantors there will be a generation after that. If liberal states are not cranking out bloodthirsty warriors with the skills, knowledge, and attributes to zealously conduct winning networking warfare, they are going to get crushed. Be a band of brothers.

(5) Unleash the hounds. Networking for national security is meaningless unless the networks are operationalized to deliver meaningful, impactful, and relevant action that serves the national interests of free peoples. Never be afraid to fight for freedom.

Power of Right

Let me end with one last story on the power of networking. I hope it will convince you that this power is real and within your grasp.

I mentioned that when the war against Ukraine broke out, we set up another ad hoc network. One day, I was connected to Phoebe Olhava, a Boston doctor who had volunteered to help Ukrainian refugees. In early 2022, she was contacted by a group trying to evacuate fifteen Ukrainian orphans. "It began when I asked you for a contact who might help with an ambulance transfer of an intubated infant in April 2022," she later wrote me. I gave her two contacts. "From there," she added, "I made other contacts who I reached out to when I heard about this particular group of orphans who had been forcibly moved from Mykolaiv to Kherson and then into Russia just before Kherson was liberated. They have safely made it to Georgia where they are staying until they can return to Mykolaiv [Ukraine]."[15] This stuff really works.

Thank You

If you stuck with *Digital Dominance* to the end, you are either an evildoer who is looking for ways to make better mischief or someone who cares about making the world a better place for the planet and the people who live on it.

If you are an evildoer, then be afraid, be very afraid that good people take seriously the advice in *Digital Dominance*. If you are good people, then thank you for caring about good people. Godspeed and fair winds.

NOTES

Prologue

1. As reported in "Taliban Says Will Respect Women's Rights, Press Freedom," Al Jazeera, August 17, 2021, https://www.aljazeera.com/news/2021/8/17/taliban-says-will-respect-womens-rights-press-freedom.

2. In May 2011, the National Directorate of Security, Afghanistan's intelligence agency, claimed to have discovered that Mujahid is really "Haji Ismail," a forty-two-year-old man from the town of Chaman in Balochistan, Pakistan. When reached by cell phone, Mujahid denied that he was in Pakistan. See George Wright, "Afghanistan: Mysterious Taliban Spokesman Finally Shows His Face," BBC News, August 17, 2021, https://www.bbc.com/news/world-asia-58250607.

3. See Emerson T. Brooking, "Before the Taliban Took Afghanistan, It Took the Internet," *New Atlanticist*, August 26, 2021, https://www.atlanticcouncil.org/blogs/new-atlanticist/before-the-taliban-took-afghanistan-it-took-the-internet/; Chris Stokel-Walker, "The Battle for Control of Afghanistan's Internet," *Wired*, September 9, 2021, https://www.wired.co.uk/article/afghanistan-taliban-internet/.

4. There are different versions of this story in the press. Here is one: "The Driver from Pakistan Stopped by to Look at the Kabul Airport and Ended Up in the USA," *World News Today*, August 27, 2021, https://www.world-today-news.com/the-driver-from-pakistan-stopped-by-to-look-at-the-kabul-airport-and-ended-up-in-the-usa/.

5. This incident is described by a number of media outlets. See, for example, Yaron Steinbuch, "Desperate Afghan Moms Throw Babies over Barbed Wire to UK Troops at Airport," *New York Post*, August 19, 2021, https://nypost.com/2021/08/19/afghan-moms-throw-babies-over-barbed-wire-to-uk-troops-at-airport/.

6. They were eventually evacuated. "Uzbekistan Completes Deportation of Hundreds of Afghan Pilots and Family," Radio Free Europe/Radio Liberty, September 13, 2021, https://www.rferl.org/a/afghanistan-pilots-uzbekistan-evacuation/31457269.html.

7. A month later, thousands were still stranded. Pauline Smolinski, "Almost 150 from the American University of Afghanistan Were Evacuated, but Thousands Still Want to Leave," CBS News, September 22, 2021, https://www.cbsnews.com/news/american-university-of-afghanistan-evacuated-but-thousands-still-want-to-leave/.

8. See "ISIS Releases Chilling Video of Apparent Execution," CBS News, February 3, 2015, https://www.youtube.com/watch?v=xzwhyfO04u8.

Chapter 1

1. Kermit Pattison, *Fossil Men: The Quest for the Oldest Skeleton and the Origins of Humankind* (New York: William Morrow, 2020), 333.

2. Networking, to be fair, was not just a Homo sapiens activity. Other humanoid species adopted similar practices of organizing familial and community activities. See Lee Berger and John Hawks, *Cave of Bones, a True Story of Discovery, Adventure, and Human Origins* (Washington, DC: National Geographic, 2023).

3. Bob Yirka, "Early Humans Gained Energy Budget by Increasing Rate of Energy Acquisition, Not Energy-Saving Adaptation," Phys.org, December 30, 2021, https://phys.org/news/2021–12-early-humans-gained-energy-acquisition.amp.

4. Steven Pinker, "Language as an Adaptation to the Cognitive Niche," in *Language Evolution*, ed. Morton H. Christiansen, and Simon Kirby (New York: Oxford University Press, 2003).

5. The social structure of primitive Papua New Guinea is described in D. K. Feil, *The Evolution of Highland Papua New Guinea Societies* (Cambridge: Cambridge University Press, 1987), 123–66.

6. Daniel Okamura, "Powering Down: Theoretical Lenses to Examine the Agency of Our Smartphones," in *Interfacing Ourselves: Living in the Digital Age*, ed. Cristina Bodinger-deUriate (New York: Routledge, 2019), 45.

7. Paul Starr, *The Creation of the Media* (New York: Basic Books, 2004), 1.

8. Starr, *The Creation of the Media*, 198–205.

9. Kris Ruijgrok, "From the Web to the Streets: Internet and Protests Under Authoritarian Regimes," *Democratization* 24, no. 3 (2017): 498–520.

10. Research concludes that social media did not "create" the Arab Spring, but online activity did act as a significant accelerant for political action. Heather Brown, Emily, Guskin, and Amy Mitchell, "The Role of Social Media in the Arab Uprisings," Pew Research Center, November 28, 2012, https://www.pewrsearch.org/journalism/2012/11/28/role-social-media-arab-uprisings/; Gadi Wolfsfeld, Elad Segev, and Tamir Sheafer, "Social Media and the Arab Spring: Politics Comes First," *International Journal of Press/Politics* 18, no. 2 (2012): 115–37; Adam Smidi and Saif Shahin, "Social Media and Social Mobilisation in the Middle East," *India Quarterly* 73, no. 2 (2017): 196–209.

11. Prasanta Kumar Dutta, "Protests Sweep Around the Globe as Israel's War in Gaza Grinds On," Reuters, November 13, 2023, https://www.reuters.com/

graphics/ISRAEL-PALESTINIANS/MAPS/movajdladpa/#protests-sweep-around-the-globe-as-israels-war-in-gaza-grinds-on.

12. For example, see the debate over treating big tech companies as monopolies and breaking them up. William Rinehart, "Breaking Up Big Tech Is Hard to Do, Innovation Depends on Large Companies' Teams and Shared Technologies," *Wall Street Journal*, July 22, 2018, https://www.wsj.com/articles/breaking-up-big-tech-is-hard-to-do-1532290123; Zachery Karabell, "Don't Break Up Big Tech," *Wired*, January 23, 2020, https://www.wired.com/story/dont-break-up-big-tech/; Robert C. O'Brien, "Breaking Up Tech Is a Gift to China, Populist Proposals Would Punish the Companies Competing with Beijing on AI and Quantum," *Wall Street Journal*, December 26, 2021, https://www.wsj.com/articles/congress-breaking-up-silicon-valley-tech-is-a-gift-to-china-tencent-baidu-bytedance-quantum-11640525284.

13. Unless cited otherwise, information in the section "Update" is derived from GlobalWebIndex, "Social," 2019, p. 12, https://www.globalwebindex.com/hubfs/Downloads/2019%20Q1%20Social%20Flagship%20Report.pdf; Brooke Auxier and Monica Anderson, "A Majority of Americans Say They Use YouTube and Facebook, While Use of Instagram, Snapchat and TikTok Is Especially Common Among Adults Under 30," Pew Research Center, April 7, 2021, https://www.pewresearch.org/internet/2021/04/07/social-media-use-in-2021/; Aleksandra Atanasova, "Gender-Specific Behaviors on Social Media and What They Mean for Online Communications," Social Media Today, November 6, 2016, https://www.socialmediatoday.com/social-networks/gender-specific-behaviors-social-media-and-what-they-mean-online-communications.

14. See, for example, Nancy R. Ahern and Brandy Mechling, "Sexting: Serious Problems for Youth," *Journal of Psychosocial Nursing and Mental Health Services* 51, no. 7 (2013), https://doi.org/10.3928/02793695–20130503–02; Wendy Moncur, Kathryn M. Orzech, and Fergus G. Neville, "Fraping, Social Norms and Online Representations of Self," *Computers in Human Behavior* 63, no. 1 (2016): 125–31.

15. Osaji Obi, "State of Internet Scams 2023," September 27, 2023, https://social-catfish.com/scamfish/state-of-internet-scams-2023/.

16. Alan Suderman, Frank Bajak, and Rodney Muhumuza, "Africa Internet Riches Plundered, Contested by China Broker," AP, October 1, 2021, https://apnews.com/article/technology-business-africa-china-uganda-5462f03bbd75b-f9724a26623295fbf0e.

17. Samuel P. L. Veissière and Moriah Stendel, "Hypernatural Monitoring: A Social Rehearsal Account of Smartphone Addiction," *Frontiers in Psychology*, February 20, 2018, https://doi.org/10.3389/fpsyg.2018.00141.

18. Belinda Luscombe, "How Ryan Kaji Became the Most Popular 10-Year-Old in the World," *Time*, November 12, 2021, https://time.com/6116624/ryan-kaji-youtube. For the evolution of the YouTube partnership revenue sharing program, see Robert Kyncl and Maany Peyvan, *Streampunks: YouTube and the Rebels Remaking Media* (New York: Harper, 2017), 20–21.

19. Benjamin Golub and Matthew O. Jackson, "Naïve Learning in Social Networks and the Wisdom of Crowds," *American Economic Journal: Microeconomics* 2, no. 1 (2010): 112–49.

20. Georgia Wells, Jeff Horwitz, and Deepa Seetharaman, "Facebook Knows Instagram Is Toxic for Teen Girls, Company Documents Show," *Wall Street Journal*, September 14, 2021, https://www.wsj.com/articles/facebook-knows-instagram-is-toxic-for-teen-girls-company-documents-show-11631620739; "Facebook Puts Instagram for Kids on Hold After Pushback," Associated Press, September 27, 2021, https://www.bostonherald.com/2021/09/27/facebook-puts-instagram-for-kids-on-hold-after-pushback/amp/.

21. See, for example, "Instagram for Kids: Setting Up Instagram Parental Control," Bark, updated August 16, 2023, https://www.bark.us/tech-guide/app-management-instagram.

22. Amanda Trejos, "Ice Bucket Challenge: 5 Things You Should Know," *USA Today*, July 3, 2017, https://www.usatoday.com/story/news/2017/07/03/ice-bucket-challenge-5-things-you-should-know/448006001/. See also Geah Pressgrove, Brooke Weberling McKeever, and S. Mo Jang, "What Is Contagious? Exploring Why Content Goes Viral on Twitter: A Case Study of the ALS Ice Bucket Challenge," *Journal of Philanthropy and Marketing* 23, no. 1 (2018), https://doi.org/10.1002/nvsm.1586; Agnes Ng May Phing and Rashad Yazdanifard, "How Does ALS Ice Bucket Challenge Achieve Its Viral Outcome Through Marketing via Social Media?" *Global Journal of Management and Business Research* 14, no. 7 (2014): 57–61.

23. There is a vast literature on this subject. See, for example, Jacques deLisle, Avery Goldstein, and Guobin Yang, eds., *The Internet, Social Media, and a Changing China* (Philadelphia: University of Pennsylvania Press, 2021).

24. Kevin Lewis, Kurt Gray, and Jens Meierhenrich, "The Structure of Online Activism," *Sociological Science*, February 18, 2014, 1–9.

25. Asma Shakir Khawaja and Asma Hussain Khan, "Media Strategy of ISIS," *Strategic Studies* 36, no. 2 (2016): 104–21.

26. Julie Uldam, "Social Media Visibility: Challenges to Activism," *Media, Culture and Society* 40, no. 1 (2017): 41–58.

27. Daniel Victor, "Woman and Homeless Man Plead Guilty in $400,000 GoFundMe Scam," *New York Times*, March 7, 2019, https://www.nytimes.com/2019/03/07/us/gofundme-homeless-scam-guilty.html.

28. Organisation of Economic Co-operation and Development, "Disinformation and Russia's War of Aggression Against Ukraine Threats and Governance Responses," November 3, 2022, https://www.oecd.org/ukraine-hub/policy-responses/disinformation-and-russia-s-war-of-aggression-against-ukraine-37186bde/.

29. Although as of this writing it is still premature to declare winners and losers in the online Ukraine war, see, for example, Peter Suciu, "Ukraine Is Winning on the Battlefield and on Social Media," *Forbes*, October 13, 2022, https://www

.forbes.com/sites/petersuciu/2022/10/13/ukraine-is-winning-on-the-battlefield-and-on-social-media/?sh=7ba722e44008.

30. Adrian Furnham, "The Brainstorming Myth," *Business Strategy Review* 11, no. 4 (2000): 21–28. See also Karen Leggett Dugosh, Paul B. Paulus, Evelyn Roland, and Huei-Chuan Yang, "Cognitive Stimulation in Brainstorming," *Journal of Personality and Social Psychology* 79, no. 5 (2000): 722–35.

31. Alan Feuer and Jason George, "Internet Fame Is Cruel Mistress for a Dancer of the Numa Numa," *New York Times*, February 6, 2005, https://www.nytimes.com/2005/02/26/nyregion/internet-fame-is-cruel-mistress-for-a-dancer-of-the-numa-numa.html.

32. On Estonian innovation, see, for example, D. Azzopardi et al., "Seizing the Productive Potential of Digital Change in Estonia," OECD Economics Department Working Papers no. 1639 (2020), https://doi.org/10.1787/999c7d5a-en.

33. Nick Heath, "How Estonia Became an E-government Powerhouse," Tech Republic, February 19, 2019, https://www.techrepublic.com/article/how-estonia-became-an-e-government-powerhouse/; Department of Economic and Social Affairs, United Nations, *E-Government Survey 2020 Digital Government in the Decade of Action for Sustainable Development* (New York: United Nations, 2020), 20.

34. Phil Tully, "Vevo Hacked via LinkedIn Phishing Campaign, over 3TB of Sensitive Data Exposed," Zero Fox, September 19, 2017, https://www.zerofox.com/blog/vevo-hacked-via-linkedin-phishing-campaign.

35. Hannah Smith and Katherine Mansted, "Weaponised Deep Fakes, National Security and Democracy," Policy Brief Report no. 28/2020, Australian Strategic Policy Institute, April 2020, p. 12.

36. "Robottrolling," NATO Strategic Communications Centre of Excellence, September 6, 2021, https://stratcomcoe.org/pdfjs/?file=/publications/download/Robotrolling_06sept2021_fin.pdf?zoom=page-fit.

37. Nguyen Phong Hoang et al., "How Great Is the Great Firewall? Measuring China's DNS Censorship," 2021, https://arxiv.org/pdf/2106.02167.pdf.

Chapter 2

1. Sarah Grant, "Watch Katy Perry 'Rise' and 'Roar' for Hillary Clinton at DNC," *Rolling Stone*, July 29, 2016, https://www.rollingstone.com/politics/politics-news/watch-katy-perry-rise-and-roar-for-hillary-clinton-at-dnc-123015/.

2. For more on the flow of foreign fighters, see Brian Dodwell, Daniel Milton, and Don Rassler, "The Caliphate's Global Workforce: An Inside Look at the Islamic State's Foreign Fighter Paper Trail," Combatting Terrorism Center, April 18, 2016.

3. Arushi Arora, Sumit Kumar Yadav, and Kavita Sharma, "Denial-of-Service (DoS) Attack and Botnet: Network Analysis, Research Tactics, and Mitigation," in *Research Anthology on Combating Denial-of-Service Attacks* (Hershey, PA: Information Resources Management Association, 2021), 49–73.

4. "Uncovering the C&C Communication Capabilities of Malicious YouTube Bot," Cycle, December 23, 2022, https://cyble.com/blog/new-youtube-bots-malware-spotted-stealing-users-sensitive-information.

5. Robert Cookson, "Google Charges for YouTube Ads Even When Viewed by Robots," *Financial Times*, September 22, 2015, https://www.ft.com/content/53ac3fd0–604e-11e5-a28b-50226830d644; John Koetsier, "YouTube Bug Cutting Creator Pay 50% or More; YouTubers Becoming Desperate," *Forbes*, November 6, 2020, https://www.forbes.com/sites/johnkoetsier/2020/11/06/youtube-bug-cutting-creator-pay-50-or-more-youtubers-becoming-desperate/?sh=7171e592753e; Catalin Cimpanu, "LoT Botnet Used in YouTube Ad Fraud Scheme," ZDNet, January 31, 2019, https://www.zdnet.com/article/iot-botnet-used-in-youtube-ad-fraud-scheme/.

6. See, for example, Scott Jasper, *Russian Cyber Operations: Coding the Boundaries of Conflict* (Washington, DC: Georgetown University Press, 2020), 35, 38–39. This, of course, is not argument that Russia is the only significant player in illicit cyber activity or that all malicious Russian cyber activity is directed by the Russian government. See, for example, Athina Karazogianni, "Blame It on the Russians: Tracking the Portrayal of Russian Hackers During Cyber Conflict Incidents," in *Violence and War in Culture and the Media*, ed. Athina Karazogianni (London: Routledge, 2011), 221–46.

7. For a robust discussion of the concept of national security, see Kim Holmes, "What Is National Security?" in *Index of U. S. Military Strength*, ed. Dakota Woods (Washington, DC: Heritage Foundation, 2015), 17–25.

8. See, for example, National Resources Council of Maine, "Climate Change Impacts on National Security," accessed October 20, 2021, https://www.nrcm.org/programs/federal/federal-climate-and-energy-issues/climate-change-impacts-on-national-security/.

9. For an explanation, see James Jay Carafano, "Welcome to Obama's War on Weather," Heritage Foundation, June 2014, https://www.heritage.org/defense/commentary/welcome-obamas-war-weather; James Jay Carafano, "National Security: Not a Good Argument for Global Warming Legislation," Heritage Foundation, October 28, 2009, https://www.heritage.org/testimony/national-security-not-good-argument-global-warming-legislation.

10. See John M. Hawkins, "The Costs of Artillery: Harassment and Interdiction Fire in the Vietnam War" (master's thesis, Texas A&M University, 2004), https://hdl.handle.net /1969.1/ETD-TAMU-2004-THESIS-H285.

11. Unless noted otherwise, quotes and information in this section are drawn from Hedy Greijdanus, Carlos Ade Matos Fernandes, Felicity Turner-Zwinkels, Ali Honari, Carla A. Roos, Hannes Rosenbusch, and Tom Postmes, "The Psychology of Online Activism and Social Movements: Relations Between Online and Offline Collective Action," *Current Opinion in Psychology* 35 (October 2020): 49–54.

12. Anne Speckhard, "The Boston Marathon Bombers: The Lethal Cocktail that Turned Troubled Youth to Terrorism," *Perspectives on Terrorism* 7, no. 3: 64–78. See also Anne Speckhard, "Recruiting from Beyond the Grave: A European

Follows Anwar al-Awlaki into ISIS," Homeland Security Today, April 28, 2020, https://www.hstoday.us.

13. Steven Lloyd Wilson, "Detecting Mass Protest Through Social Media," *Journal of Social Media in Society* 6, no. 2 (2017): 5–21.

14. Aaron Noland, "Social Media Activists: Analyzing the Relationship Between Online Activism and Offline Attitudes and Behaviors," *Journal of Social Media in Society* 6, no. 2 (2017): 26–47.

15. Noland, "Social Media Activists," 26; Katherine White and John Peloza, "Self-Benefit Versus Other-Benefit Marketing Appeals: Their Effectiveness in Generating Charitable Support," *Journal of Marketing* 73, no. 4 (2009): 109–24; James Surowiecki, "What Happened to the Ice Bucket Challenge?" *New Yorker*, July 18, 2016, https://www.newyorker.com/magazine/2016/07/25/als-and-the-ice-bucket-challenge.

16. Alina Mungiu-Pippidi and Igor Munteanu, "Moldova's 'Twitter Revolution,'" *Journal of Democracy* 20, no. 3 (2009): 136–42.

17. John Arquilla, a security analyst at RAND, is one of the pioneers in the field of understanding network war and its broader applications. See, for example, John Arquilla and David Ronfeldt, eds., *Networks and Netwars: The Future of Terror, Crime, and Militancy* (Santa Monica, CA: Rand, 2001).

18. For fulsome discourse on the center of gravity concept, see Antulio J. Echavarria II, *Clausewitz's Center of Gravity: Changing Our Warfighting Doctrine—Again!* (Carlisle Barracks, PA: US Army War College, 2002).

19. John J. Morrow Jr., "To Command the Sky: The Battle for Air Superiority over Germany, 1942–1944," *Technology and Culture* 3, no. 4 (1993): 958–60.

20. Jon Tetsuro Sumida, *Decoding Clausewitz: A New Approach to* On War (Lawrence: University of Kansas Press, 2008), 163–64; Roger Parkinson, *Clausewitz* (New York: Stein and Day, 1979), 308–9; quotes from Carl von Clausewitz in *On War*, ed. and trans. Michael Howard and Peter Paret Princeton: Princeton University Press, 1976), 566.

21. Surowiecki, "What Happened to the Ice Bucket Challenge?"

22. Described in Chuck Remsberg, "Staying 'Left of Bang': What to See and Do to Prevent Deadly Attacks," Force Science News, February 1, 2015, https://www.forcescience.org/2015/02/staying-left-of-bang-what-to-see-and-do-to-prevent-deadly-attacks/.

Chapter 3

1. For more on this see, see James Jay Carafano and Paul Rosenzweig, *Winning the Long War: Lessons from the Cold War for Defeating Terrorism and Preserving Freedom* (Washington, DC: Heritage Foundation, 2005), 22–50; James Jay Carafano, "Omens of a Hollow Military," *Heritage Foundation*, September 4, 2013, https://www.heritage.org/defense/commentary/omens-hollow-military; James Jay Carafano, "How to Halt the Slide Towards a Hollow U.S. Military," October 2, 2014, https://www.heritage.org/budget-and-spending/commentary/how-halt-the-slide-towards-hollow-us-military.

2. The efforts of the transition team were documented in Christopher Liddell, Daniel Kroese, and Clark Campbell, *Romney Readiness Project: Retrospective & Lessons Learned* (Lexington, KY: R2P, 2013).

3. For more, see Chris King, "Inside the Trump Transition Team with James Carafano," *St. Louis Dispatch*, May 10, 2017, https://www.stlamerican.com/inside-the-trump-transition-team-with-james-carafano/article_0435ac1e-35f1–11e7–8872–17f46e42cc64.html.

4. "Transcript of Donald Trump's Speech on National Security in Philadelphia," *The Hill*, September 7, 2016, https://thehill.com/blogs/pundits-blog/campaign/294817-transcript-of-donald-trumps-speech-on-national-security-in.

5. For an assessment, see James Jay Carafano, Luke Coffey, Nile Gardiner, Walter Lohman, Terry Miller, and Thomas Spoehr, "Preparing the U.S. National Security Strategy for 2020 and Beyond," Heritage Foundation, May 23, 2019, https://www.heritage.org/defense/report/preparing-the-us-national-security-strategy-2020-and-beyond.

6. Hermina Nedelescu, "Anti-Science, Mistrust, and Anxiety in the Orthodox World," Berkley Center for Religion, Peace and World Affairs, August 26, 2021, https://berkleycenter.georgetown.edu/responses/anti-science-mistrust-and-anxiety-in-the-orthodox-world; "Vaccine Related Conversations on Facebook: A Comparison of Romania and Moldova," Beacon Project, September 20, 2021, https://www.iribeaconproject.org/our-work-analysis-and-insights/2021–09–20/vaccine-related-conversations-facebook-compairson-romania.

7. "Vaccine Related Conversations on Facebook."

8. Jeff Jarvis, *What Would Google Do?* (New York: Harper, 2009).

9. Global Engagement Center, "GEC Special Report: Pillars of Russia's Disinformation and Propaganda Ecosystem," U.S. State Department, August 2020, p. 19; "Minds Besieged: Digital Warfare Against the American Electorate," Omelas, pp. 6–12, https://www.omelas.io/mb-report. The SCF occasionally makes news in the United States when its articles are reposted by prominent individuals. See, for example, "Marjorie Taylor Greene's Russian Propaganda Post Sparks Backlash," *Newsweek*, November 30, 2023, https://www.newsweek.com/marjorie-taylor-greene-russia-ukraine-tweet-1848237.

10. See, James Jay Carafano, "The Truth About the America First Movement," National Interest, July 7, 2016, https://nationalinterest.org/feature/the-truth-about-the-america-first-movement-16887.

11. Florence Passy and Marco Giugni, "Social Networks and Individual Perceptions: Explaining Differential Participation in Social Movements," *Sociological Forum* 16 (2001): 123–53.

12. Brett Schaefer, "How to Make the State Department More Effective at Implementing U.S. Foreign Policy," Heritage Foundation, April 20, 2016, https://www.heritage.org/political-process/report/how-make-the-state-department-more-effective-implementing-us-foreign. The secretary got fired before he could implement his reforms, but that wasn't Brett's fault. His job was to just sell the idea in the first place. That he did.

13. See, for example, Paul Castelloe and Joshua Prokopy, "Recruiting Participants for Community Practice Interventions," *Merging Community Practice Theory and Social Movement Theory* 45, no. 4 (2008): 31–48.

14. Aaron Smith and Monica Anderson, "Social Media Use in 2018," Pew Research Center, March 1, 2018, https://www.pewresearch.org/internet/2018/03/01/social-media-use-in-2018/.

15. Maurice Dawson, Max Lieble, and Adewale Adeboje, "Open Source Intelligence: Performing Data Mining and Link Analysis to Track Terrorist Activities," in *Information Technology—New Generations. Advances in Intelligent Systems and Computing*, ed. Shahram Latifi (Zurich: Springer International, 2018), 159–63.

Chapter 4

1. Christopher Hamner, "Why Do Soldiers Fight?" *Historically Speaking* 13, no. 1 (2012): 10–12.

2. This section is adapted from James Jay Carafano, "War, Peace, Philanthropy: Bringing the Spirit of America to Combat Zones," *Philanthropy* (Summer 2015), https://www.philanthropyroundtable.org/philanthropy-magazine/article/summer-2015-war-peace-philanthropy.

3. See, for example, Almoatazbillah Hassan, "The Value Proposition Concept in Marketing: How Customers Perceive the Value Delivered by Firms—A Study of Customer Perspectives on Supermarkets," *International Journal of Marketing Studies* 4, no. 3 (2012): 68–87; Adrian Payne, Pennie Frow, and Andreas Eggert, "The Customer Value Proposition: Evolution, Development, and Application in Marketing," *Journal of the Academy of Marketing Science* 45 (2017): 467–89; Adrian Payn, Pennie Frow, Lena Steinhoff, and Andreas Eggert, "Toward a Comprehensive Framework of Value Proposition Development: From Strategy to Implementation," *Industrial Marketing Management* 87 (May 2020): 244–55.

4. Wong Sze Wan, Omkar Dastane, Nurhizam Safir Mohad Satar, and Mohamad Yusnorizam Ma'Arif, "What WeChat Can Learn from WhatsApp: Customer Value Proposition Development for Mobile Social Networking (MSN) Apps: A Case Study Approach," *Journal of Theoretical and Applied Information Technology* 97, no. 4 (2019): 1103, 1116.

5. Natalie Wolchover, "Are flat-Earthers Being Serious?" *Live Science*, May 30, 2017, https://www.livescience.com/24310-flat-earth-belief.html. See also J. Eric Oliver and Thomas J. Wood, "Conspiracy Theories and the Paranoid Style(s) of Mass Opinion," *American Journal of Political Science* 58, no. 4 (2014): 952–66; Danny Lewis, "The Curious History of the International Flat Earth Society," *Smithsonian Magazine*, January 29, 2016, https://www.smithsonianmag.com/smart-news/curious-history-international-flat-earth-society-180957969/.

6. Katherine Thomson, "Jonah Hill's Twitter Nightmare (VIDEO)," *HuffPost*, December 6, 2017, https://www.huffpost.com/entry/jonah-hills-twitter-night_n_231312; Amy Wilkinson, "Jonah Hill's Twitter Imposter: Found!" MTV News, September 14, 2009, http://www.mtv.com/news/2550615/jonah-hills-twitter-impostor-found/.

Chapter 5

1. This section is adopted from James Jay Carafano, "War, Peace, Philanthropy: Bringing the Spirit of America to Combat Zones," *Philanthropy Magazine* (Summer 2015), https://www.philanthropyroundtable.org/magazine/summer-2015-war-peace-philanthropy/.

2. Quotes are from "Celebrating Incredible Women: Laos Buffalo Diary," Spirit of America, March 13, 2019, https://spiritofamerica.org/blog/celebrating-incredible-women-laos-buffalo-dairy.

3. Quotes from The Infantry School, *Infantry in Battle* (Washington, DC: The Infantry Journal Incorporated, 1939), 139.

4. See, for example, David Silver, *The Social Network Business Plan* (New York: Willey, 2009); Ali Tayebi, "Planning Activism: Using Social Media to Claim Marginalized Citizens' Right to the City," *Cities* 32 (2013): 51–59.

5. Vicki J. Rast and Dylan Lee Lehrke, *Interagency Paralysis: Stagnation in Bosnia and Kosovo* (Washington, DC: Project on National Security Reform, 2011), 384–450.

6. See, for example, Chad Storlie, "Manage Uncertainty with Commander's Intent," *Harvard Business Review*, November 30, 2010, https://hbr.org/2010/11/dont-play-golf-in-a-football-g.

7. See, for example, Antoine Bourguilleau, Phillippe Lepinard, and Natalia Wojtowiz, "Wargames for Training Future Managers," *Management and Data Science* 5, no. 1 (December 2020), https://management-datascience.org/articles/14547/.

8. Described in Micah Zenko, *Red Team: How to Succeed by Thinking Like the Enemy* (New York: Basic Books, 2020).

9. "DHS 2.0—Rethinking the Department of Homeland Security," Center for Strategic and International Studies–The Heritage Foundation, December 13, 2004, https://csis-website-prod.s3.amazonaws.com/s3fs-public/legacy_files/files/media/csis/pubs/041213_dhsv2.pdf.

10. "Department of Homeland Security: The Road Ahead," Committee on Homeland Security and Governmental Affairs, January 25, 2005, https://www.govinfo.gov/content/pkg/CHRG-109shrg20169/html/CHRG-109shrg20169.htm.

11. Ashley Mae Orcutt, "State of Internet Scams 2021," Social Catfish, accessed December 3, 2021, https://socialcatfish.com/blog/state-of-internet-scams-2021/.

12. This list is adapted from "Good Cyber Hygiene Habits to Help You Stay Safe Online," Kaspersky, accessed December 1, 2021, https://usa.kaspersky.com/resource-center/preemptive-safety/cyber-hygiene-habits.

13. NordPass, accessed December 28, 2021, https://nordpass.com/most-common-passwords-list/.

14. Pierluigi Paganini, "The Most Common Social Engineering Attacks [Updated 2020]," InfoSec, August 6, 2020, https://resources.infosecinstitute.com/topic/common-social-engineering-attacks/.

15. D. Wallace and J. Costello, "Eye in the Sky: Understanding the Mental Health of Unmanned Aerial Vehicle Operators," *Journal of Military and Veterans Health* 25, no. 3 (2017): 36–39.

16. See, for example, Rosó Baltà-Salvador, Noelia Olmedo-Torre, Marta Peña, and Ana-Inés Renta-Davids, "Academic and Emotional Effects of Online Learning During the COVID-19 Pandemic on Engineering Students," *Education and Information Technologies* 26 (2021): S7407–7434.

17. See, for example, Sally Hardy and Ben Thomas, "Mental and Physical Health Comorbidity: Political Imperatives and Practice Implications," *International Journal of Mental Health Nursing* 21, no. 3 (2012): 289–98.

18. See William J. Broad, *The Science of Yoga: The Risks and the Rewards* (New York: Simon & Schuster, 2012).

19. Jennifer W. Adkins, "Mental Health in the Workplace: The Value of Rest," National Alliance of Mental Illness, August 28, 28, 2017, https://www.nami.org/Blogs/NAMI-Blog/August-2017/Mental-Health-in-the-Workplace-The-Value-of-Rest.

20. Chandra Kant, Nlawade Rajesh, and Pradhan Seema, "Stress Relieving Techniques for Organizational Stressors," *International Journal of Research in Commerce & Management* 7, no. 3 (2016): 93–98.

21. See, for example, Alan Ewert and Yun Chang, "Levels of Nature and Stress Response," *Behavioral Science* 8, no. 5 (2018): 49.

Chapter 6

1. For more on the relationship between wartime learning and the postwar private sector, as well as the adaptive skills of soldiers, see James Jay Carafano, *GI Ingenuity: Improvisation, Technology and Winning World War II* (Boulder: Lynne Rienner, 2006).

2. Robert W. Goldfarb, "The Elephant in the Room Is 86," *New York Times*, March 23, 2017, https://www.nytimes.com/2017/03/23/well/live/the-elephant-in-the-room-is-86.html.

3. Goldfarb, "The Elephant in the Room."

4. Daniel R. Goethel, Sean M. Lucey, Aaron M. Berger, Sarah K. Gaichas, Melissa A. Karp, Patrick D. Lynch, John F. Walter III, Jonathan J. Deroba, Shana Miller, and Michael J. Wilberg, "Closing the Feedback Loop: On Stakeholder Participation in Management Strategy Evaluation," *Canadian Journal of Fisheries and Aquatic Sciences* 3 (December 2018), https://doi.org/10.1139/cjfas-2018–0162@cjfas-mse.issue01.

5. See James Jay Carafano, "Measuring Military Power," Heritage Foundation, September 2, 2014, https://www.heritage.org/defense/commentary/measuring-military-power.

6. See Daniel Nelson, *Frederick W. Taylor and the Rise of Scientific Management* (Madison: University of Wisconsin Press, 1980).

7. Discussed in Carafano, *GI Ingenuity*.

8. James Jay Carafano, *Wiki at War: Conflict in a Socially Networked World* (College Station: Texas A&M University Press, 2012), 247–48.

9. For a discussion of 360 assessments, see Jack Zenger and Joseph Folkman, "What Makes a 360-Degree Review Successful?" *Harvard Business Review*,

December 23, 2020, https://hbr.org/2020/12/what-makes-a-360-degree-review-successful.

10. Susanne Salem-Schatz, Diana Ordin, and Brian Mittman, "Guide to the After Action Review," Center for Evidence-Based Management, October 2010, https://www.yumpu.com/en/document/view/35728512/guide-to-the-after-action-review-queri; David A Garvin, *Learning In Action: A Guide to Putting the Learning Organization to Work* (Boston: Harvard Business School Press, 2000), 106–16.

11. Self-assessments are only as reliable as the quality of the leader making them. See Kenneth J. Meier and Laurence J. O'Toole Jr., "I Think (I Am Doing Well), Therefore I Am: Assessing the Validity of Administrators' Self-Assessments of Performance," *International Public Management Journal* 16, no. 1 (2013): 1–27.

12. Carafano, *Wiki at War*, 5–7.

13. Brad Martin, Thomas Manacapilli, James C. Crowley, Joseph Adams, Michael G. Shanley, Paul Steinberg, and Dave Stebbins, "Assessment of Joint Improvised Explosive Device Defeat Organization (JIEDDO) Training Activity," RAND, 2013, https://www.rand.org/content/dam/rand/pubs/research_reports/RR400/RR421/RAND_RR421.pdf; Peter Carey and Nancy Youssef, "JIEDDO, the Manhattan Project that Bombed," Center for Public Integrity, March 27, 2011, https://publicintegrity.org/national-security/jieddo-the-manhattan-project-that-bombed/; Ronald F. A. Woodaman, Andrew G. Loerch, and Kathryn B. Laskey, "A Decision Analytic Approach for Measuring the Value of Counter-IED Solutions at the Joint Improvised Explosive Device Defeat Organization," George Mason University, 2010, http://c4i.gmu.edu/eventsInfo/reviews/2010/papers/Woodaman_Valuation_of_Counter-IED_Solutions.pdf.

Chapter 7

1. Quotes and information in this section are from "How the Feds Took Down the Silk Road Drug Wonderland," *Wired*, November 18, 2013, https://www.wired.com/2013/11/silk-road/.

2. This section is adapted from James Carafano and Richard Weitz, "Complex Systems Analysis—A Necessary Tool for Homeland Security," Heritage Foundation, April 16, 2009, https://www.heritage.org/homeland-security/report/complex-systems-analysis-necessary-tool-homeland-security.

3. James Jay Carafano, "Talking Through Disasters: The Federal Role in Emergency Communications," Heritage Foundation, July 17, 2006, https://www.heritage.org/defense/report/talking-through-disasters-the-federal-role-emergency-communications.

4. See, for example, Committee on Network Science for Future Army Applications Board on Army Science and Technology, Division on Engineering and Physical Sciences, *Network Science* (Washington, DC: National Academies Press, 2005).

5. "23 Free Social Network Analysis Tools [as of 2021]," RankRed, August 11, 2021, https://www.rankred.com/free-social-network-analysis-tools/.

6. SurveyMonkey, accessed November 2, 2021, https://www.surveymonkey.com/.
7. See China Transparency Project, https://www.heritage.org/china-transparency-project. See also their annual report: Walter Lohman and Justin Rhee, eds., *The 2021 Chinese Transparency Report* (Washington, DC: Heritage Foundation, 2021).
8. AWS Data Exchange, https://aws.amazon.com/data-exchange/.
9. Tyler Sonnemaker, "Law Enforcement Agencies Are Using a Legal Loophole to Buy Up Personal Data Exposed by Hackers," Business Insider, July 8, 2020, https://www.businessinsider.com/police-buying-hacked-data-bypassing-legal-processes-2020–7.
10. David Collier, "Understanding Process Tracing," *Political Science and Politics* 44, no. 4 (2011): 823.
11. See, for example, François Fouss, *Algorithms and Models for Network Data and Link Analysis* (Cambridge: Cambridge University Press, 2016).
12. Microsoft Analysis Services, https://help.tableau.com/current/pro/desktop/en-us/examples_msas.htm.
13. Martin Saveski, Eric Chu, Soroush Vosoughi, and Deb Roy, "Human Atlas: A Tool for Mapping Social Networks," WWW'16 Companion, April 11–15, 2016, http://dx.doi.org/10.1145/2872518.2890552.
14. Jonathan Gips and Alex Pentland, "Mapping Human Networks," Fourth Annual IEEE International Conference on Pervasive Computing and Communications (PERCOM'06) (New York: IEEE, 2006), 10.
15. See, for example, Headquarters, Department of the Army, Intelligence Preparation of the Battlefield, March 2019, https://home.army.mil/wood/application/files/8915/5751/8365/ATP_2–01.3_Intelligence_Preparation_of_the_Battlefield.pdf.

Chapter 8

1. Bob Dreyfus, "Sebastian Gorka, the West Wing's Phony Foreign-Policy Guru," *Rolling Stone*, August 10, 2017, https://www.rollingstone.com/politics/politics-features/sebastian-gorka-the-west-wings-phony-foreign-policy-guru-129772/.
2. Eli Stokols, Bryan Bender, and Michael Crowley, "The Husband-and-Wife Team Driving Trump's National Security Policy," Politico, February 13, 2017, https://www.politico.com/story/2017/02/trump-national-security-gorka-234950.
3. See, for example, Kara Frederick, "Combating Big Tech's Totalitarianism: A Road Map," Heritage Foundation, February 7, 2022, https://www.heritage.org/technology/report/combating-big-techs-totalitarianism-road-map.
4. Robert Hart, "Parler's Popularity Plummets as Data Reveals Little Appetite for Returning 'Free Speech' App Favored by Conservatives," *Forbes*, June 2, 2021, https://www.forbes.com/sites/roberthart/2021/06/02/parlers-popularity-plummets-as-data-reveals-little-appetite-for-returning-free-speech-app-favored-by-conservatives/?sh=4686c4e95e13.
5. Jack Stubbs, "Payment Sent—Travel Giant CWT Pays $4.5 Million Ransom to

Cyber Criminals," Reuters, July 31, 2020, https://www.reuters.com/article/us-cyber-cwt-ransom/payment-sent-travel-giant-cwt-pays-4–5-million-ransom-to-cyber-criminals-idUSKCN24W25W.

6. See Robert Gildea, *Fighters in the Shadows: A New History of the French Resistance* (Boston: Harvard University Press, 2015).

7. Collin Mulliner, Nico Golde, and Jean-Pierre Seifert, "SMS of Death: From Analyzing to Attacking Mobile Phones on a Large Scale" USENIX Security Symposium, 2011, https://www.usenix.org/legacy/event/sec11/tech/full_papers/Mulliner.pdf.

8. See, for example, Department of the Army, *Army Support for Military Deception, FM 3–13.4* (Washington, DC: Department of the Army, 2019).

9. R. W. Burns, "Deception, Technology and the D-Day Invasion," *Engineering Science and Education Journal* 4, no. 2 (1995): 81–88.

10. Hemank Lamba, Shashank Srikanth, Dheeraj Reddy Pailla, Shwetanshu Singh, Karandeep Singh Juneja, and Ponnurangam Kumaraguru, "Driving the Last Mile: Characterizing and Understanding Distracted Driving Posts on Social Networks," *Proceedings of the Fourteenth International AAAI Conference on Web and Social Media* (2020): 393–403.

11. Bethania Palma, "Photos Spark Flurry of Rhetoric About 'Whips' at the Border," Snopes, September 29, 2021, https://www.snopes.com/news/2021/09/29/border-whipping-photographs/; Kevin Johnson and Rebecca Morin, "DHS Inspector General Declines Review of Horse-Mounted Agents in Haitian Migrant Confrontation," *USA Today*, November 16, 2021, https://www.usatoday.com/story/news/politics/2021/11/16/dhs-watchdog-mounted-agents-haitian-migrants/8637319002/.

12. Gary King and Jennifer Pan, "How the Chinese Government Fabricates Social Media Posts for Strategic Distraction, Not Engaged Argument," *American Political Science Review* 111, no. 3 (2017): 484–501.

13. See, for example, Julian Landavazo, "The Effectiveness of Allied Airborne Units on D-Day" (master's thesis, University of New Mexico, 2011), https://digitalre-pository.unm.edu/cgi/viewcontent.cgi?article=1042&context=hist_etds.

14. Reuters Staff, "NYPD Analysis Opposed WTC Command Center Site," Reuters, January 26, 2008, https://www.reuters.com/article/us-newyork-security-giuliani/nypd-analysis-opposed-wtc-command-center-site-paper-idUSB21557420080126.

15. Patrick Traynor, Michael Lin, Machigar Ongtang, Vikhyath Rao, Trent Jaeger, Patrick McDaniel, and Thomas La Porta, "On Cellular Botnets: Measuring the Impact of Malicious Devices on a Cellular Network Core," *Proceedings of the 16th ACM Conference on Computer and Communications Security* (November 2009): 223–34.

16. Described in Elizabeth Sayles and Rick Beyer, *The Ghost Army of World War II: How One Top-Secret Unit Deceived the Enemy with Inflatable Tanks, Sound Effects, and Other Audacious Fakery* (Princeton: Princeton Architectural Press, 2015).

17. For an introduction to business continuity planning and emergency management planning, see Mark Sauter and James Carafano, *Homeland Security*, 3rd ed. (New York: McGraw Hill, 2013).

18. ConservativeStack.Com, accessed December 28, 2021, https://conservativestack.com/.

19. See, for example, Terri Howard, "Social Media and Business Continuity Planning," *Risk Management*, May 8, 2013, http://www.rmmagazine.com/articles/article/2013/05/08/-Social-Media-and-Business-Continuity-Planning; "Following the Feed: Business Continuity in the Age of Social Media," Continuity Insights, March 7, 2019, https://continuityinsights.com/following-the-feed-business-continuity-in-the-age-of-social-media/.

Chapter 9

1. The deputy national security adviser's remembrances during the transition and the White House are described in K. T. McFarland, *Revolution: Trump, Washington and "We the People"* (Franklin, TN: Post Hill Press, 2020), 133–203.

2. To read the strategy for yourself, see "The National Security Strategy of the United States," White House, 2017, https://trumpwhitehouse.archives.gov/wp-content/uploads/2017/12/NSS-Final-12–18–2017–0905.pdf.

3. For analysis, see James Jay Carafano et al., "Preparing the U.S. National Security Strategy for 2020 and Beyond," Heritage Foundation, May 23, 2019, https://www.heritage.org/defense/report/preparing-the-us-national-security-strategy-2020-and-beyond.

4. See James Carafano, Walter Lohman, Thomas Spoehr, Luke Coffey, David R. Shedd, and Nile Gardiner, "5 Reasons H. R. McMaster Is the Right Leader for a Tough President," Daily Signal, August 3, 2017, https://www.dailysignal.com/2017/08/03/5-reasons-h-r-mcmaster-right-leader-tough-president/; James Jay Carafano, "Trump's Growing Collection of National Security Advisors," National Interest, March 21, 2020, https://nationalinterest.org/feature/trump%E2%80%99s-growing-collection-national-security-council-advisors-135652.

5. For an explanation, see James Jay Carafano, "Lose Your Cool and Die—5 Firefight Films You Need to Watch Now," National Interest, May 14, 2021, https://nationalinterest.org/blog/buzz/lose-your-cool-and-die%E2%80%945-firefight-films-you-need-watch-now-185201.

6. The doctrine and controversies surrounding are considered in Walter LaFeber, "The Rise and Fall of Colin Powell and the Powell Doctrine," *Political Science Quarterly* 124, no. 1 (2009): 71–93.

7. Perry Biddiscombe, *The Last Nazis: SS Werewolf Guerrilla Resistance in Europe 1944–1947* (London: Tempus, 2004).

8. For instance, see the development of the army's assessment under General George C. Marshall in Charles E. Kilpatrick, *An Unknown Future and a Doubtful Present: Writing the Victory Plan of 1941* (Washington, DC: Center of Military History, 1992).

9. See James Jay Carafano, *Brutal War* (Boulder: Lynne Rienner, 2021), 63.

10. Mary-Hunter McDonnell and Brayden G. King, "Keeping Up Appearances: Reputational Threat and Impression Management After Social Movement Boycotts," *Administrative Science Quarterly* 58, no. 3 (2013): 387–419.

11. For a discussion of the concept, see J. Boone Bartholomees, "Theory of Victory," *Parameters* (Summer 2008): 25–36.

12. Carafano, *Brutal War*, 236.

13. See, for example, Ben Smith, "Big Tech Has Crushed the News Business. That's About to Change," *New York Times*, May 10, 2020, https://www.nytimes.com/2020/05/10/business/media/big-tech-has-crushed-the-news-business-thats-about-to-change.html.

14. This section and quotes are drawn from Brian Martin, "Online Onslaught: Internet-Based Methods for Attacking and Defending Citizens' Organisations," *First Monday* 17, no. 12 (2012).

15. See, for example, Yan Sun and Yuhong Liu, "Security of Online Reputation Systems: The Evolution of Attacks and Defenses," *IEEE Signal Processing Magazine* 29, no. 2 (2012): 87—97.

16. Sam Haysom, "Microsoft Is Launching New Technology to Fight Deepfakes, a Video Authenticator Is on Its Way," Mashable, September 2, 2020, https://mashable.com/article/microsoft-technology-fight-deepfakes.

Chapter 10

1. James Jay Carafano and Thomas Spoehr, "Don't Let Academia Destroy Military History," National Interest, March 19, 2021, https://nationalinterest.org/blog/buzz/don%E2%80%99t-let-academia-destroy-military-history-179724.

2. See Russell D. Budhite and W. M. Christopher Hamel, "War for Peace: The Question of an American Preventive War Against the Soviet Union, 1945–1955," *Diplomatic History* 14, no. 3 (1990): 367–84.

3. For a discussion of the concept, see Thomas Hurka, "Proportionality in the Morality of War," *Philosophy & Public Affairs* 33, no. 1 (2005): 34–66.

4. Section adapted from James Jay Carafano et al., "The Great Eastern Japan Earthquake: Assessing Disaster Response and Lessons for the U.S.," Heritage Foundation, May 25, 2011, https://www.heritage.org/asia/report/the-great-eastern-japan-earthquake-assessing-disaster-response-and-lessons-the-us.

5. See James Jay Carafano and Richard Weitz, "EMP Attacks—What the U.S. Must Do Now," Heritage Foundation, November 17, 2010, https://thf_media.s3.amazonaws.com/2010/pdf/bg2491.pdf.

6. Nathan Hodge, "E-Bomb Awareness Day: Grab Your Tin Foil Hat," *Wired*, March 25, 2018, https://www.wired.com/2010/03/e-bomb-awareness-day-grab-your-tinfoil-hat/.

7. For more on this debate, see Sean Lawson and Michael K. Middleton, "Cyber Pearl Harbor: Analogy, Fear, and the Framing of Cyber Security Threats in the United States, 1991–2016," *First Monday* 24, no. 3 (2019), https://doi.org/10.5210/fm.v24i3.9623.

8. See Nathan A. Jennings, "Nuclear Weapons and the Korean War: A Precarious Beginning for the Tradition of Non-Use," *Small Wars Journal*, November 4, 2014, https://smallwarsjournal.com/jrnl/art/nuclear-weapons-and-the-korean-war-a-precarious-beginning-for-the-tradition-of-non-use; Peter Hayes and Nina Tannenwald, "Nixing Nukes in Vietnam," *Bulletin of Atomic Scientists* (May/June 2003): 52–59.

9. See, for example, Gabriel D. Lourenço, "Can Hackers Bring Down a Government? In Belarus Activists Are Trying; Experts Answer," Olhar Digital, September 9, 2021, https://olhardigital.com.br/en/2021/09/08/seguranca/hackers-derrubar-governo/.

10. See, for example, Kim Murphy, "Lawsuits Threaten to Drain the Life out of Hate Groups," *Los Angeles Times*, August 22, 2000, https://www.latimes.com/archives/la-xpm-2000-aug-22-mn-8425-story.html.

11. Togzhan Kassenov, "The Exploitation of the Global Financial Systems for Weapons of Mass Destruction (WMD) Proliferation," testimony before the House of Representatives Financial Services Committee, March 4, 2020, https://carnegieendowment.org/2020/03/04/exploitation-of-global-financial-systems-for-weapons-of-mass-destruction-wmd-proliferation-pub-81221.

12. See an extensive history discussed in Kelly M. Greenhill, *Weapons of Mass Migration: Forced Displacement, Coercion, and Foreign Policy* (Ithaca: Cornell Studies in Security Affairs, 2016).

13. Greenhill, *Weapons of Mass Migration*, 3.

14. See, for example, Gabriel Gavin, "More EU Countries Threaten to Close Russian Borders amid Migrant Surge," *Politico*, December 4, 2023, https://www.politico.eu/article/eu-baltic-countries-threaten-close-russian-borders-kremlin-migrant-surge-finland.

15. See, for example, "Africa's Population Will Double by 2050," *Economist*, March 26, 2020, https://www.economist.com/special-report/2020/03/26/africas-population-will-double-by-2050.

16. Gil Murciano, "Unpacking the Global Campaign to Delegitimize Israel: Drawing the Line Between Criticism of Israel and Denying Its Legitimacy," Stiftung Wissenschaft und Politik, June 22, 2020.

17. See, for example, Ben White, "Delegitimizing Solidarity: Israel Smears Palestine Advocacy as Anti-Semitic," *Journal of Palestine Studies* 49, no. 2 (2020): 65–79.

18. See, for example, Joanne Smith Finley, "Why Scholars and Activists Increasingly Fear a Uyghur Genocide in Xinjiang," *Journal of Genocide Research* 23, no. 3 (2021): 348–70.

19. Uğur Ümit Üngör, "Forum: Mass Violence in Syria," *Journal of Genocide Research* 25, no. 1 (2021): 84–88.

Chapter 11

1. Samantha Aschieris, "Xi, Putin 'Want to Create Their Own New Rules,' Asian Studies Expert Says," Daily Signal, September 13, 2023, https://www

.dailysignal.com/2023/09/13/xi-putin-want-create-their-own-new-rules-asian-studies-expert-says.

2. See Nathan Ruser, "Documenting Xinjiang's Detention System," Australian Strategic Policy Institute, September 2020.

3. Mike Pompeo, "Determination of the Secretary of State on Atrocities in Xinjiang," Press Statement, US State Department, January 19, 2021, https://2017–2021 .state.gov/determination-of-the-secretary-of-state-on-atrocities-in-xinjiang/ index.html.

4. For dissenting views see, for example, Michael J. Mazarr, "This Is Not a Great-Power Competition, Why the Term Doesn't Capture Today's Reality," *Foreign Affairs*, May 29, 2019, https://www.foreignaffairs.com/articles/2019–05–29/ not-great-power-competition; Daniel H. Nexon, "Against Great Power Competition, the U.S. Should Not Confuse Means for Ends," *Foreign Affairs*, February 15, 2021, https://www.foreignaffairs.com/articles/united-states/2021–02–15/against-great-power-competition; Matej Kandrík, "The Case Against the Concept of Great Power Competition" Strategy Bridge, June 30, 2021, https://thestrategybridge.org/the-bridge/2021/6/30/the-case-against-the-concept-of-great-power-competition?format=amp.

5. See, for example, Congressional Research Service, "Renewed Great Power Competition: Implications for Defense—Issues for Congress," CRS Report R43838, November 17, 2021.

6. Quoted in Dean Cheng, Walter Lohman, James Carafano, and Riley Walters, "Assessing Beijing's Power: A Blueprint for the U.S. Response to China over the Next Decades," Heritage Foundation, February 10, 2020, https://www .heritage.org/asia/report/assessing-beijings-power-blueprint-the-us-response-china-over-the-next-decades.

7. See, for example, the Department of Defense report described in Bill Gertz, "Pentagon Details China Info War on U.S.," *Washington Times*, November 3, 2021, https://www.washingtontimes.com/news/2021/nov/3/pentagon-details-china-info-war-us/.

8. These differences are described in James Jay Carafano, "Maybe 'Civilizations' Aren't the Problem," National Interest, July 7, 2018, https://nation-alinterest.org/feature/maybe-civilizations-aren%E2%80%99t-problem-25187?page=0%2C1.

9. Carafano, "Maybe 'Civilizations' Aren't the Problem."

10. Described in Dean Cheng, *Cyber Drago: Inside China's Information Warfare and Cyber Operations* (Santa Barbara: ABC-CLIO, 2016).

11. This section is adapted from James Jay Carafano and Chad Wolf, "China's Damaging Influence and Exploitation of U.S. Colleges and Universities," Heritage Foundation, March 22, 2021, https://www.heritage.org/asia/ commentary/chinas-damaging-influence-and-exploitation-us-colleges-and-universities.

12. Janet Lorin and Brandon Kochkodin, "Harvard Leads U.S. Colleges that Received $1 Billion from China," Bloomberg, February 6, 2020, https://www

.bnnbloomberg.ca/harvard-leads-u-s-colleges-that-received-1-billion-from-china-1.1385895.

13. Wolf and Carafano, "China's Damaging Influence."

14. James Jay Carafano et al., "Winning the New Cold War: A Plan for Countering China," Heritage Foundation, March 28, 2023, https://www.heritage.org/asia/report/winning-the-new-cold-war-plan-countering-china.

15. This section is adapted from James Jay Carafano, "How NATO Can Avoid the Death Spiral on Europe's Frontier," Heritage Foundation, November 9, 2021, https://www.heritage.org/global-politics/commentary/how-nato-can-avoid-the-death-spiral-europes-frontier.

16. Bureau of Counterterrorism, "Country Reports on Terrorism 2019: Iran," US Department of State, 2019, https://www.state.gov/reports/country-reports-on-terrorism-2019/iran/.

17. See, for example, Annie Fixler and Divjot Bawa, "Iran's Social Engineering Capabilities Mature," Foundation for the Defense of Democracies, July 23, 2021, https://www.fdd.org/analysis/2021/07/23/irans-social-engineering-capabilities-mature/.

18. Ivana Kottasová and Sara Mazloumsaki, "The 'Internet as We Know It' Is off in Iran. Here's Why This Shutdown Is Different," CNN, November 19, 2019, https://www.cnn.com/2019/11/19/middleeast/iran-internet-shutdown-intl/index.html. See also "Internet Disrupted in Iran amid Fuel Protests in Multiple Cities," NetBlocks, November 15, 2019, https://netblocks.org/reports/internet-disrupted-in-iran-amid-fuel-protests-in-multiple-cities-pA25L18b.

19. James Andrew Lewis, "Iran and Cyber Power," Center for Strategic and International Studies, June 25, 2019, https://www.csis.org/analysis/iran-and-cyber-power.

20. See, for example, "Publicly Reported Iranian Cyber Actions in 2019," Center for Strategic and International Studies, https://www.csis.org/programs/technology-policy-program/publicly-reported-iranian-cyber-actions-2019.

21. Bruce Klingner, "North Korean Cyberattacks: A Dangerous and Evolving Threat," Heritage Foundation, September 2, 2021, https://www.heritage.org/asia/report/north-korean-cyberattacks-dangerous-and-evolving-threat.

22. Public-Private Analytic Exchange Program, "Commodification of Cyber Capabilities, a Grand Cyber Arms Bazaar," Department of Homeland Security, 2019, p. 5.

23. Steve Morgan, "Humans on the Internet Will Triple from 2015 to 2022 and Hit 6 Billion," CyberCrime Magazine, July 18, 2019, https://cybersecurityventures.com/how-many-internet-users-will-the-world-have-in-2022-and-in-2030/.

24. Joseph Johnson, "Cyber Crime: Number of Breaches and Records Exposed 2005–2020," Statista, March 3, 2021, https://www.statista.com/statistics/273550/data-breaches-recorded-in-the-united-states-by-number-of-breaches-and-records-exposed/.

25. "Advanced Persistent Threat Actors Targeting U.S. Think Tanks," Joint Cyber

Security Advisory, December 1, 2020, https://us-cert.cisa.gov/sites/default/files/publications/AA20–336A-APT_Actors_Targeting_US_ThinkTanks.pdf.

26. "Evidence Suggests that the U.S. Loses Hundreds of Billions to Cybercrime, Possibly as Much as 1 % to 4 % of GDP Annually," National Institute of Standards and Technology, May 27, 2020, https://www.nist.gov/news-events/news/2020/05/evidence-suggests-us-loses-hundreds-billions-cybercrime-possibly-much-1–4.

27. "Channeling the Tide, Protecting Democracies amid a Flood of Corrosive Capital," Center for International Private Enterprise, 2019, pp. 2–3. This report principally focuses on China and Russia.

28. Christopher Walker and Jessica Ludwig, "From 'Soft Power' to 'Sharp Power': Rising Authoritarian Influence in the Democratic World," in *Sharp Power: Rising Authoritarian Influence* (Washington, DC: National Endowment for Democracy, 2017).

29. James Jay Carafano, "International Organizations Are the Devil's Playground of Great Power Competition," National Interest, May 15, 2020, https://nationalinterest.org/feature/international-organizations-are devil%E2%80%99s-playground-great-power-competition-154706?page=0%2C1.

30. See, for example, the collection of materials on the website of the Japan Foreign Ministry at https://www.mofa.go.jp/policy/page25e_000278.html. See also US State Department, "A Free and Open Indo-Pacific: A Shared Vision," US State Department, November 4, 2019; David Brewster, "A Free and Open Indo-Pacific," Lowry Institute, March 7, 2018, https://www.lowyinstitute.org/the-interpreter/free-and-open-indo-pacific-and-what-it-means-australia.

31. See Global Engagement Center, US State Department, https://www.state.gov/bureaus-offices/under-secretary-for-public-diplomacy-and-public-affairs/global-engagement-center/.

32. For an example of the center's work, see Global Engagement Center, "Pillars of Russia's Disinformation and Propaganda Ecosystem," US State Department, August 2020.

33. Matthew C. Weed, "Global Engagement Center: Background and Issues," CRS Insights, August 4, 2017.

34. Office of the Inspector General, "Audit of Global Engagement Center Federal Assistance Award Management and Monitoring," US Department of State, April 2020, Highlights.

35. See, for example, James Jay Carafano, "Staying Ahead of the Islamist Terrorist Threat: Assessing Future Domestic Counterterrorism Measures," testimony before Subcommittee on National Security Committee on Oversight and Government Reform, United States House of Representatives, June 23, 2016, pp. 1–7; Human Factors/Behavioral Sciences Division, Science and Technology Directorate, "The Organizational Dynamics of Far-Right Hate Groups in the United States: Comparing Violent to Non-Violent Organizations," US Department of Homeland Security, December 2011.

36. See, for example, the case of US-India cooperation. James Jay Carafano, "A Strong U.S.-India Partnership Is in Our Strategic Interest," National Interest, December 12, 2017, https://nationalinterest.org/feature/strong-us-india-partnership-our-strategic-interest-23624.

37. John Seaman, "Towards a More China-Centered Global Economy? Implications for Chinese Power in the Age of Hybrid Threats," European Centre of Excellence for Countering Hybrid Threats, November 2021, p. 7.

38. Gulizar Haciyakupoglu, "China's Social Credit System: Questions on the Current Status, Role of Data and Surveillance, and Influence Outside of China," NATO Strategic Communications Centre of Excellence, 2021.

39. Piret Pernik, "Cyber Deterrence: A Case Study on Estonia's Policies and Practice," European Centre of Excellence for Countering Hybrid Threats, October 2021, p. 11.

40. See, for example, James Jay Carafano, "China's Sanctions on U.S. Officials Are an Attempt to Scare the World," July 20, 20202, https://scnow.com/opinion/columnists/james-jay-carafano-china-s-sanctions-on-u-s-officials-are-an-attempt-to-scare/article_73d754ee-ebd7–5a5e-a031–77ea050ed3d1.html.

41. James Jay Carafano, "It's Not NATO, but Quad Group Can Get Results in Asia," Heritage Foundation, October 19, 2020, https://www.heritage.org/global-politics/commentary/its-not-nato-quad-group-can-get-results-asia.

42. See James Jay Carafano and Anthony Kim, "The US–South Korea Alliance Is a Historic Success (and Could Get Even Better)," 19FortyFive, September 8, 2021, https://www.19fortyfive.com/2021/09/the-us-south-korea-alliance-is-a-historic-success-and-could-get-even-better/.

43. See, for example, James Jay Carafano, "Time for a Collective Defense in the Middle East," Heritage Foundation, September 23, 2019, https://www.heritage.org/middle-east/commentary/time-collective-defense-the-middle-east.

44. Matthew Cohen, Charles Freilich, and Gabi Siboni, "Israel and Cyberspace: Unique Threat and Response," *International Studies Perspectives* 17, no. 3 (2016): 307–21; Islam Alhalwany, "Israel Is Becoming a Cybersecurity Guarantor in the Middle East. Here's How," MENASource, November 18, 2021, https://www.atlanticcouncil.org/blogs/menasource/israel-is-becoming-a-cybersecurity-guarantor-in-the-middle-east-heres-how/.

45. James Jay Carafano, "Why Small States Matter to Big Powers," National Interest, August 13, 2018, https://www.heritage.org/defense/commentary/why-small-states-matter-big-powers.

46. This section is adapted from James Jay Carafano, "Grading Dictatorships," National Interest, January 23, 2016, https://nationalinterest.org/feature/grading-dictators-top-task-the-next-president-14996?amp.

Chapter 12

1. Kim D. Reimann, "A View from the Top: International Politics, Norms and the Worldwide Growth of NGOs," *International Studies Quarterly* 50, no. 1 (2006): 45–67.

2. Bureau of Democracy, Human Rights, and Labor, "Non-Governmental Organizations (NGOs) in the United States," US State Department, January 20, 2021.
3. Willard Foxton, "If Silk Road Was a Legitimate Startup, It Would Be Worth $2.4 Billion," *Daily Telegraph*, October 4, 2013, https://www.businessinsider.com/silk-road-valuation-worth-2-or-3-billion-2013–10#:~:text=Silk%20Road%20Valuation%3A%20Worth%20%242%20or%203%20Billion.
4. Ellie Silverman, Ian Shapira, Tom Jackman, and John Woodrow Cox, "Spencer, Kessler, Cantwell and Other White Supremacists Found Liable in Deadly Unite the Right Rally," *Washington Post*, November 23, 2021, https://www.washingtonpost.com/dc-md-va/2021/11/23/charlottesville-verdict-live-updates/.
5. James Jay Carafano, "Forget Coronavirus: 7 Films Explain Why We Still Worry About the Zombie Apocalypse," Heritage Foundation, May 24, 2020, https://www.heritage.org/defense/commentary/forget-coronavirus-7-films-explain-why-we-still-worry-about-the-zombie.
6. James Jay Carafano and Richard Weitz, "EMP Attacks: What the U.S. Must Do Now," Heritage Foundation, November 17, 2010, https://static.heritage.org/2010/pdf/bg2491.pdf.
7. Vanda Felbab-Brown, "The Key Trends to Watch This Year on Nonstate Armed Actors," Brookings Institution, January 15, 2021, https://www.brookings.edu/blog/order-from-chaos/2021/01/15/the-key-trends-to-watch-this-year-on-nonstate-armed-actors/amp/.
8. *"Little Green Men": A Primer on Modern Russian. Unconventional Warfare, Ukraine 2013–2014* (Fort Bragg, NC: United States Army Special Operations Command, n.d.).
9. Conor M. Kennedy and Andrew S. Erickson, "China's Third Sea Force, the People's Armed Forces Maritime Militia: Tethered to the PLA," US Naval Academy, March 2017, https://digital-commons.usnwc.edu/cgi/viewcontent.cgi?article=1000&context=cmsi-maritime-reports.
10. Priscilla Alvarez, "Report: Human Smugglers Increasingly Use Facebook to Advertise Services on the US-Mexico Border," CNN, April 16, 2021, https://www.cnn.com/2021/04/16/politics/facebook-border-human-smuggling/index.html.
11. Megan Gibson, "Can a Social Media Campaign Really #BringBackOurGirls?" *Time*, May 7, 2014, https://time.com/90693/bringbackourgirls-nigeria-boko-haram/.
12. See, for example, James G. McGann and R. Kent Weaver, *Think Tanks and Civil Societies: Catalysts for Ideas and Action* (New York: Routledge, 2002).
13. "Southern Poverty Law Center (SPLC)," Influence Watch, accessed November 26, 2021, https://www.influencewatch.org/non-profit/southern-poverty-law-center-splc/.
14. Debbie Elliott, "After Allegations of Toxic Culture, Southern Poverty Law Center Tries to Move Forward," NPR, April 17, 2019, https://www.npr.org/2019/04/17/713887174/after-allegations-of-toxic-culture-southern-poverty-law-center-tries-to-move-for.

15. Jessica Prol Smith, "The Southern Poverty Law Center Is a Hate-Based Scam that Nearly Caused Me to Be Murdered," *USA Today*, August 17, 2019, https://www.usatoday.com/story/opinion/2019/08/17/southern-poverty-law-center-hate-groups-scam-column/2022301001/.

16. Christopher Dickey, "It's Not Easy to Fight Extremists if You're Called One: Why the SPLC Had to Pay Maajid Nawaz $3.375 Million," *Daily Beast*, June 20, 2918, https://www.thedailybeast.com/its-not-easy-to-fight-extremists-if-youre-called-one-why-the-sclc-had-to-pay-maajid-nawaz-dollar3375-million.

17. See, for example, Kelley Beaucar Vlahos, "New Report Details $174 Million in Foreign Funding to D.C. Think Tanks," Responsible Statecraft, January 29, 2020, https://responsiblestatecraft.org/2020/01/29/new-report-details-174-million-in-foreign-funding-to-d-c-think-tanks/.

18. This section is adapted from James Carafano, "War, Peace, Philanthropy: Bringing the Spirit of America to Combat Zones," *Philanthropy Magazine*, Summer 2015, https://www.philanthropyroundtable.org/magazine/summer-2015-war-peace-philanthropy/.

19. "Eastern Europe," CIPE, accessed December 1, 2021, https://www.cipe.org/projects/eastern-europe/.

20. Anthony Kim, "Ivanka Trump's Global Development and Prosperity Initiative Empowers Women," Heritage Foundation, February 21, 2020, https://www.heritage.org/international-economies/commentary/ivanka-trumps-global-development-and-prosperity-initiative; "United States Government Women, Peace and Security (WPS) Congressional Report," Office of the White House, June 2021, https://www.whitehouse.gov/wp-content/uploads/2021/07/USG_Women_Peace_Security_WPS_Congressional_Report_FINAL6.30.2021-Updated-July-16.pdf.

21. Influence Watch, "Council on American Islamic Relations," accessed November 29, 2021, https://www.influencewatch.org/non-profit/council-on-american-islamic-relations-cair/.

22. Leila Fadel, "Muslim Civil Rights Leader Accused Of Harassment, Misconduct, NPR, April 15, 2021, https://www.npr.org/2021/04/15/984572867/muslim-civil-rights-leader-accused-of-harassment-misconduct.

Chapter 13

1. Fox Business Staff, "Puzder on Options for Investors Opposing Corporate America's Woke Agenda," Fox Business, April 20, 2021, https://www.foxbusiness.com/politics/corporate-america-imposing-woke-agenda-unknowingly-on-investors-andy-puzder.

2. Martin J. Medhurst, "Eisenhower's Rhetorical Leadership: An Interpretation," in *Eisenhower's War of Words* (East Lansing: Michigan State University Press, 1994), 294.

3. For example, see Catherine Flick, "Why Google's Censored Search Engine for China Is an Ethical Minefield," *Independent*, August 8, 2018, https://www

.independent.co.uk/life-style/gadgets-and-tech/features/china-great-firewall-google-online-internet-censorship-project-dragonfly-a8480016.html; Sung Wook and Aziz Douaib, "Google vs. China's 'Great Firewall': Ethical Implications for Free Speech and Sovereignty," *Technology in Society* 34, no. 2 (2012): 174–81.

4. Liam Tung, "Google Employee Protest: Now Google Backs off Pentagon Drone AI Project," ZDNet, June 4, 2018, https://www.zdnet.com/article/google-employee-protests-now-google-backs-off-pentagon-drone-ai-project/.

5. Vivek Ramaswamy, *Woke Inc.: Inside Corporate America's Social Justice Scam* (New York: Center Street, 2021), 175.

6. Louise Matsakisse, "Does TikTok Really Pose a Risk to US National Security?" *Wired*, July 17, 2020, https://www.wired.com/story/tiktok-ban-us-national-security-risk/.

7. Kara Frederick, "Testimony Before Subcommittee on Communications and Technology, United States House of Representatives," December 1, 2021, 1.

8. Robert E. Bluey, "Big Tech's Conservative Censorship Inescapable and Irrefutable," Heritage Foundation, September 23, 2021, https://www.heritage.org/technology/commentary/big-techs-conservative-censorship-inescapable-and-irrefutable.

9. "Buying Influence: How China Manipulates Facebook and Twitter," *New York Times*, December 21, 2021, https://www.deccanherald.com/international/world-news-politics/buying-influence-how-china-manipulates-facebook-and-twitter-1063089.html.

10. "Buying Influence."

11. Steve Stecklow and Jeffrey Dastin, "Amazon Partnered with China Propaganda Arm,' Reuters, December 17, 2021, https://www.reuters.com/world/china/amazon-partnered-with-china-propaganda-arm-win-beijings-favor-document-shows-2021–12–17/.

12. See, for example, Brian Fung, "'Near-Perfect Market Intelligence': Why a House Report Says Big Tech Monopolies Are Uniquely Powerful," CNN, October 10, 2020, https://www.cnn.com/2020/10/10/tech/apple-amazon-facebook-amazon-monopoly-data/index.html; Matt Rosoff, "This Week Showed How the Big Tech Antitrust Campaign Is Totally Misguided," CNBC, June 30, 2020, https://www.cnbc.com/2021/06/30/op-ed-antitrust-crusade-against-big-tech-is-misguided.html; Daren Bakst and Gabriella Beaumont-Smith, "A Conservative Guide to the Antitrust and Big Tech Debate," Heritage Foundation, December 1, 2020, https://www.heritage.org/technology/report/conservative-guide-the-antitrust-and-big-tech-debate.

13. Katie Canales, "Tim Cook Says Apple Has a 'Responsibility' to Do Business Everywhere, even in China despite Its Human Rights Issues," Business Insider, November 10, 2021, https://www.businessinsider.com/apple-china-tim-cook-responsibility-world-peace-world-trade-2021–11.

14. This section is adapted from James Jay Carafano, "More Free Markets Will

Mean Fewer Wars," Heritage Foundation, January 29, 2012, https://www
.heritage.org/defense/commentary/more-free-markets-will-mean-fewer-wars.

15. This section is adapted from James Jay Carafano, "The Smart Way to Put China
in Its Place," Heritage Foundation, May 5, 2020, https://www.heritage.org/
asia/commentary/the-smart-way-put-china-its-place.

16. Frederico Bartels, "China Defense Spending in Context: How Under-Reporting
and Differing Standards and Economies Distort the Picture," Heritage Foundation,
March 25, 2020, https://www.heritage.org/asia/report/chinas-defense-budget-
context-how-under-reporting-and-differing-standards-and-economies.

17. See, for example, Valerie Insinna, "Silicon Valley Warns the Pentagon: 'Time Is
Running Out,' After Years of Research and Development Efforts, Tech Startups
Want to See the Pentagon Commit to Real Production Work," BreakingDefense,
December 21, 2021.

18. See, for example, Aaron Mehta, "US Warns Against 'Protectionism' with New EU
Defense Agreement," Munich Security Conference, February 2, 2018, https://
www.defensenews.com/smr/munich-security-forum/2018/02/14/us-warns-
against-protectionism-with-new-eu-defense-agreement/; Jakob Hanke Vela,
"EU Readies Response to Biden's 'Buy American' Pitch," Politico, May 4, 2021,
https://www.politico.eu/article/eu-response-joe-biden-buy-american-us/.

19. For criticism of the Jones Act, see, for example, Colin Grabow, Inu Manak, and
Daniel J. Ikenson, "The Jones Act: A Burden America Can No Longer Bear," Cato
Institute, June 28, 2018, https://www.cato.org/publications/policy-analysis/
jones-act-burden-america-can-no-longer-bear.

20. David John Marotta, "What Is Rent-Seeking Behavior?" *Forbes*, February 24,
2013, https://www.forbes.com/sites/davidmarotta/2013/02/24/what-is-rent-
seeking-behavior/?sh=cf81f96658af.

21. Vivek Ramaswamy, "Woke, Inc: Why I'm Blowing Whistle on How Corporate
America Is Poisoning Society," *New York Post*, June 21, 2021, https://nypost.
com/2021/06/21/woke-inc-why-im-blowing-whistle-on-how-corporate-
america-is-poisoning-society/.

22. See, for example, John H. Ohly, *Industrialists in Olive Drab: The Emergency
Operations of Private Industries During World War II* (Washington, DC: Center
of Military History, 2015).

23. Klon Kitchen, "Section 230—Mend It, Don't End It," Heritage Foundation,
October 27, 2020, https://www.heritage.org/technology/report/section-
230-mend-it-dont-end-it.

24. Frederick, Testimony Before Subcommittee on Communications and
Technology, 6.

25. Press Release, "Governor Ron DeSantis Signs Bill to Stop the Censorship of
Floridians by Big Tech," State of Florida, May 24, 2021, https://www.flgov.
com/2021/05/24/governor-ron-desantis-signs-bill-to-stop-the-censorship-of-
floridians-by-big-tech/.

26. "APP Launches Online Tool to Track Big Tech's Political Influence," American

Principles Project, July 19, 2021, https://americanprinciplesproject.org/media/app-launches-online-tool-track-big-tech-political-influence/.

27. Jerry Bowyer and Charles Bowyer, "How Conservatives Can Combat 'Woke' Shareholders," *National Review*, May 27, 2020, https://www.nationalreview.com/2020/05/how-conservatives-can-combat-woke-shareholders/.

Chapter 14

1. John Venable, "The F-35A Fighter Is the Most Dominant and Lethal Multi-Role Weapons System in the World: Now Is the Time to Ramp Up Production," Heritage Foundation, May 14, 2019, https://www.heritage.org/defense/report/the-f-35a-fighter-the-most-dominant-and-lethal-multi-role-weapons-system-the-world.

2. Quotes in this paragraph are from F. S. Norman, Brendan Gill, and Thomas Meeham, "Portable Robot," in *The 60s: The Story of a Decade, the New Yorker*, ed. Henry Finder (New York: Modern Library, 2016), 332.

3. Kate Ashford and John Schmidt, "What Is Cryptocurrency?" *Forbes*, December 9, 2021, https://www.forbes.com/advisor/investing/what-is-cryptocurrency/.

4. Sean Williams, "This Is Really Happening: Microsoft Is Developing Blockchain ID Within Its Authenticator App; Microsoft Wants to Put You in Charge of Your Digital Identity," Motley Fool, February 16, 2018, https://www.fool.com/investing/2018/02/16/this-is-really-happening-microsoft-is-developing-b.aspx.

5. Alex Simons, "Microsoft's 5 Guiding Principles for Decentralized Identities," Microsoft, October 6, 2021, https://www.microsoft.com/security/blog/2021/10/06/microsofts-5-guiding-principles-for-decentralized-identities/.

6. Michael B. Greenwald, "The New Era of Digital Asset Foreign Policy," Belfer Center for Science and International Affairs, Harvard Kennedy School, July 20, 2021, https://www.belfercenter.org/publication/new-era-digital-asset-foreign-policy.

7. Statista, "Volume of Data/Information Created, Captured, Copied, and Consumed Worldwide from 2010 to 2025," accessed November 29, 2021, https://www.statista.com/statistics/871513/worldwide-data-created/.

8. James Jay Carafano, *Wiki at War: Conflict in a Socially Networked World* (College Station: Texas A&M University Press, 2011), 251–52.

9. See, for example, Ajey Lele, "Virtual Reality and Its Military Utility," *Journal of Ambient Intelligence and Humanized Computing* 4, no. 1 (2013): 17–26.

10. Statista, "Video Game Industry—Statistics & Facts," November 19, 2021, https://www.statista.com/topics/868/video-games/#:~:text=Video%20games%20are%20a%20billion,over%2077%20billion%20U.S.%20dollars.

11. Thomas Barrabi, "Meta Shareholders Rage at 'Tone-Deaf' Mark Zuckerberg's Metaverse Push," *New York Post*, October 31, 2022, https://nypost.com/2022/10/31/big-meta-shareholders-rage-at-tone-deaf-mark-zuckerbergs-metaverse-push/?utm_campaign=SocialFlow&utm_medium=SocialFlow&utm_source=NYPTwitter.

12. Linah Aburahmah, Hajar AlRawi, Yamamah Izz, and Liyakathunisa Syed, "On-line Social Gaming and Social Networking," *Procedia Computer Science* 82 (2016): 72.

13. Jaikumar Vijayan, "The NSA Tracks World of Warcraft and Other Online Games for Terrorist Clues," *Computerworld*, December 9, 2013, https://www .computerworld.com/article/2486632/the-nsa-tracks-world-of-warcraft-and-other-online-games-for-terrorist-clues.html.

14. Marco Quiroz-Gutierrez, "Digital Land Is Selling for Millions as People Scramble to Snatch Up Virtual Real Estate in the Metaverse—and It Could Be a Multitrillion-Dollar Opportunity," *Fortune*, December 3, 2021, https://fortune. com/2021/12/03/metaverse-interest-spikes-digital-real-estate-prices/.

15. Karen Schrier, "Using Augmented Reality Games to Teach 21st Century Skills," in *Proceedings of SIGGRAPH '06* (New York: ACM, 2006), https://dl.acm.org/ doi/10.1145/1179295.1179311.

16. Fortune Business Insights, "The Global Internet of Things (IoT) Market Is Projected to Grow from $381.30 Billion in 2021 to $1,854.76 Billion in 2028 at a CAGR of 25.4% in Forecast Period," accessed December 1, 2021, http:// www.fortunebusinessinsights.com/industry-reports/internet-of-things-iot-market-100307.

17. Lu Tan and Neng Wang, "Future Internet: The Internet of Things," in *Proceedings of the 3rd International Conference on Advanced Computer Theory and Engineering (ICACTE)* (New York: IEEE, 2010).

18. Katherine Atha, Jason Callahan, John Chen, Jessica Drun, Ed Francis, Kieran Green, Brian Lafferty, Joe McReynolds, James Mulvenon, Benjamin Rosen, and Emily Walz, "China's Smart Cities Development," U.S.-China Economic and Security Review Commission, January 2020.

19. Louis Coetzee and Johan Eksteen, "The Internet of Things—Promise for the Future? An Introduction," in *2011 IST-Africa Conference Proceedings* (New York: IEEE, 2011), 1–9.

20. See, for example, Mohammad Nasar, "Web 3.0: A Review and Its Future," *International Journal of Computer Applications* 185, no. 10 (2023): 41–44.

21. Stephen Baker, "The Dangerous Cure for 'AI Bias," Reason, October 10, 2022, https://reason.com/volokh/2022/10/10/stealth-quotas/#more-8206618.

22. This section is adapted from an upcoming research paper from the Heritage Foundation on the future of work. See also James Jay Carafano, "Future Computing and Cutting-Edge National Security," Heritage Foundation, July 5, 2007, https://www.heritage.org/defense/report/future-computing-and-cutting-edge-national-security; "Why Artificial Intelligence Might Not Win a War," Heritage Foundation, May 16, 2021, https://www.heritage.org/ technology/commentary/why-artificial-intelligence-might-not-win-war.

23. "IDC Forecasts Strong 12.3% Growth for AI Market in 2020 Amidst Challenging Circumstance," BusinessWire, August 4, 2020, https://www.businesswire.com/ news/home/20200804005617/en/IDC-Forecasts-Strong-12.3-Growth-for-AI-Market-in-2020-Amidst-Challenging-Circumstances.

24. See Thomas W. Malone, *Superminds: The Surprising Power of People and Computers Thinking Together* (Cambridge, MA: MIT Press, 2018).

25. Wyatt Hoffman, "Making AI Work for Cyber Defense," Center for Security and Emerging Technology, December 2021.

26. Klon Kitchen, "Quantum Science and National Security: A Primer for Policymakers," Heritage Foundation, February 5, 2019, https://www.heritage.org/technology/report/quantum-science-and-national-security-primer-policymakers.

27. See, for example, Evan R. MacQuarrie, Christoph Simon, Stephanie Simmons, and Elicia Maine, "The Emerging Commercial Landscape of Quantum Computing," *Nature Reviews Physics* 2 (2020): 596–98.

28. Adan Siegal, "When China Rules the Web: Technology in Service of the State," *Foreign Affairs*, August 13, 2018, 10; Daniel Garisto, "China Is Pulling Ahead in Global Quantum Race, New Studies Suggest," *Scientific American*, July 15, 2021, https://www.scientificamerican.com/article/china-is-pulling-ahead-in-global-quantum-race-new-studies-suggest/; John Prisco, "China: The Quantum Competition We Can't Ignore," *Forbes*, January 21, 2021, https://www.forbes.com/sites/forbestechcouncil/2021/01/21/china-the-quantum-competition-we-cant-ignore/?sh=64bc61175d19; Elsa B. Kania and John K. Costello, "Quantum Hegemony? China's Ambitions and the Challenge to U.S. Innovation Leadership," Center for a New American Security, September 2018, https://s3.amazonaws.com/files.cnas.org/documents/CNASReport-Quantum-Tech_FINAL.pdf?mtime=20180912133406.

29. Tom Garlinghouse, "Quantum Computing: Opening New Realms of Possibilities," Princeton University, January 21, 2020, https://www.princeton.edu/news/2020/01/21/quantum-computing-opening-new-realms-possibilities.

30. See, for example, Daitian Li, Tony W. Tong, and Yangao Xiao, "Is China Emerging as the Global Leader in AI?" *Harvard Business Review*, February 18, 2021, https://hbr.org/2021/02/is-china-emerging-as-the-global-leader-in-ai.

31. Kitchen, "Quantum Science and National Security."

32. Statista, "The Global Augmented Reality (AR), Virtual Reality (VR), and Mixed Reality (MR) Market Is Forecast to Reach 30.7 Billion U.S. Dollars in 2021, Rising to Close to 300 Billion U.S. Dollars by 2024," accessed November 23, 2021, https://www.statista.com/statistics/591181/global-augmented-virtual-reality-market-size/#:~:text=The%20global%20augmented%20reality%20(AR,billion%20U.S.%20dollars%20by%202024.

33. Paul Matthews, "The Future of VR and AR," IEEE, https://www.computer.org/publications/tech-news/trends/the-future-of-vr-and-arr.

34. Brian J. Park, Stephen J. Hunt, Charles Martin, Gregory J. Nadolski, Bradford J. Wood, and Terence P. Gade, "Augmented and Mixed Reality: Technologies for Enhancing the Future of IR," *Journal of Vascular and Interventional Radiology* 31, no. 7 (2020): 1074–82.

35. "Facebook to Hire 10,000 Workers in Europe to Build 'Metaverse,'" Al Jazeera, October 18, 2021, https://www.aljazeera.com/amp/economy/2021/10/18/facebook-to-hire-10000-workers-in-europe-to-build-metaverse.

36. Alexandra Kelly, "The Air Force Is Looking to Develop a Gaming Environment to Help Expedite Decision-Making in Battle Scenarios," NextGov, December 28, 2021, https://www.nextgov.com/emerging-tech/2021/12/air-force-expands-ai-training-simulated-battle-plans/360214/.

Epilogue

1. Anders S. G. Andrae and Tomas Edler, "On Global Electricity Usage of Communication Technology: Trends to 2030," *Challenges* 6, no. 1 (2015): 117.
2. Justine Calma, "The Environmental Impact of the AI Revolution Is Starting to Come into Focus," The Verge, October 10, 2023, https://www.theverge.com/2023/10/10/23911059/ai-climate-impact-google-openai-chatgpt-energy.
3. Jean-François Mercure, Pablo Salas, A. Foley, Unnada Chewpreecha, Hector Pollitt, P. B. Holden, and N. R. Edwards, "The Dynamics of Technology Diffusion and the Impacts of Climate Policy Instruments in the Decarbonisation of the Global Electricity Sector," *Energy Policy* 73 (2014): 686–700.
4. Jeng-Fung Chen, "Forecasting Monthly Electricity Demands: An Application of Neural Networks Trained by Heuristic Algorithms," *Information* 8, no. 1 (2017): 31.
5. This section is adapted from James Jay Carafano and Jack Spencer, "Protecting the Planet While Great Powers Compete," Heritage Foundation, January 11, 2021, https://www.heritage.org/environment/commentary/protecting-the-planet-while-great-powers-compete.
6. David E. Bloom and Jeffrey G. Williamson, "Demographic Transitions and Economic Miracles in Emerging Asia," *World Bank Economic Review* 12, no. 3 (1998): 419–55.
7. See, for example, Gordon H. Hanson, "The Economic Consequences of the International Migration of Labor," *Annual Review of Economics* 1 (2009): 179–208.
8. One immigration policy that seeks to address this issue is "merit-based" immigration, a framework already adopted by several industrialized nations including Canada and Australia. See, for example, Masud Chand and Rosalie L. Tung, "Skilled immigration to Fill Talent Gaps: A Comparison of the Immigration Policies of the United States, Canada, and Australia," *Journal of International Business Policy* 2 (2019): 333–55.
9. Maria Krysan, Kristin A. Moore, and Nicholas Zill, "Identifying Successful Families: An Overview of Constructs and Selected Measures," US Department of Health and Human Services, May 10, 1990, https://aspe.hhs.gov/reports/identifying-successful-families-overview-constructs-selected-measures-0.
10. Luca Braghieri, Ro'ee Levy, and Alexey Makarin, "Social Media and Mental Health," *American Economic Review* 112, no. 11 (2022): 3660.
11. Scott Zipperle, "'Digital Babies' and the Culture of Lifestyle over a Culture of Life," Heritage Foundation, July 11, 2022, https://www.heritage.org/marriage-and-family/commentary/digital-babies-and-the-culture-lifestyle-over-culture-life.

12. Max Smeets, *No Shortcuts: Why States Struggle to Develop a Military Cyber Force* (New York: Oxford University Press, 1922), 7.

13. This section is adapted from James Jay Carafano, "Thinking the Future," *Whitehead Journal of Diplomacy and International Relations* (Summer/Fall 2009): 27–37.

14. See, for example, Rami Ali and David Luther, "Scenario Planning: Strategy, Steps and Practical Examples," Brainyard, May 14, 2020, https://www.netsuite .com/portal/business-benchmark-brainyard/industries/articles/cfo-central/ scenario-planning.shtml; "Horizon Scanning: A Practitioner's Guide," Institute of Risk Management, accessed December 18, 2021, https://www.theirm.org/ media/7423/horizon-scanning_final2–1.pdf; "Delphi Method," RAND, accessed December 28, 2021, https://www.rand.org/topics/delphi-method.html.

15. You can read about their story at "15 UA Minors Extracted from Russia," Borderlands International, January 21, 2023, https://borderlandsinternational .org/2023/01/21/15-ua-minors-rescued-from-russia.

INDEX

civil-military fusion, 193
Clausewitz, Carl von, 55–56, 157
climate change, 266–67
Colby, Eldridge, 203
collateral damage, 182
complex systems, 133–36
computational power, 249–50
consumers, 242
conversation mode, 24–25, 75, 93–94, 122
corporatism, 238
correlation *vs.* causation, 119
corrosive capital, 202
cost imposition, 170–71
countermeasures, 156–57. *See also* defensive tactics
counterproliferation, 183
The Creation of the Media (Starr), 18
criminal networks, 217
crisis decisions, 100–101
criticality, 79
Crutchfield, Leslie R., 114
CSSAs (Chinese Students and Scholars Associations), 195
culminating points, 56, 163–64
Cunningham, Michael, 27
Curtis, Lisa, 215–16
cybersecurity: cyberattacks, 20–21, 28, 32–34, 153, 179–80; data security, 248, 253; defensive, 172; fundamentals, 43–47; hygiene, 109–10; Iranian offensive capabilities, 199–201; quantum computing, 258–59

Dancing with Qubits: How Quantum Computing Works and How it can Change the World (Sutor), 258
data collection, 115–25, 138–40, 249, 252–53, 255–56
data security, 248, 253
decentralization, of data, 253
decentralized leadership, 106–7
deception, 153–55
decision making, 99–101

defeat, 169–72, 176
defense, 57
Defense Intelligence Agency (DIA), 38–39
defensive tactics: against destruction, 182–83; against disruption, 148–58; feedback loops, 123–24; mapping, 141–43; risk assessment, 79–80, 219; technology, 172; types of, 57–59; value propositions, 93, 168
degradation, 155–57. *See also* offensive tactics
delaying tactics, 152–53. *See also* offensive tactics
deliberate planning, 99–100
denial of service, 151–52
denial-of-service attacks, 252
destruction, 176–85
deterrence, 171
DIA (Defense Intelligence Agency), 38–39
Dictators at War and Peace (Weeks), 209–10
digital identity, 248
diplomacy, 206–11
direct approaches, 169
direct tactics, 57
disinformation, 153–55, 197
disruption, 60
dissuasion, 171
distraction, 153–55. *See also* offensive tactics
distributed networks, 168
Duffy, Brooke Erin, 251
Duyvesteyn, Isabelle, 166

economic freedom, 234–36, 239–40
education, 242–43, 268–69
80/20 rule, 22–23
Eisenhower, Dwight D., 68, 98, 230–31
electromagnetic pulse (EMP) strikes, 178–79
emergent technologies, 247–63

Kitchen, Klon, 258
knowledge, 104–5

language evolution, 17
leadership, 102–7, 112
legislative reforms, 241
Lettow, Paul, 105
Lewis, Jim, 199
LikeWar: The Weaponization of Social Media (Singer and Brooking), 32–33, 217
Limited War: The Challenge to American Strategy (Osgood), 222
linear thinking, 28–29, 31, 35, 75, 204
link analysis software, 78
lobbying, 85–86
Lovejoy, Owen, 16

machine learning, 254–57
malicious networks, 217–20
Malone, Thomas, 255
malware, 153
maneuvering, 170
Mansoor, Peter, 173–75, 185
mapping, 60, 131–43. *See also* feedback
marketing, 88–91
Marshall, George C., 98
mass disruption, 184–85
mass migration, 184
Mattis, Jim, 87–88
McAvoy, Jenny, 226
McGann, Jim, 265
McMaster, H.R., 161–63, 171, 215, 216
measures, 156–57. *See also* offensive tactics
Medhurst, Martin, 230
mental health, 26, 110–11, 269–70
messaging, 72–74
metrics, 116–21
Middle East, 208–9
middle states, 209–11
military history, 173–85
military-industrial complex, 230–31
missile defenses, 182–83

mission statements, 68–72, 102
Moldova, 197
monopolies, 233, 239

NATO (North Atlantic Treaty Organization), 197, 206–8
network attacks, 20–21, 28, 32–34, 153, 179–80. *See also* cybersecurity
network building: recruiting, 58–59, 68–78, 88–89, 168; retention, 86–94, 168; risk mitigation, 79–80
network destruction, 60–61
network dominators, 28–31, 35, 77, 83–84, 204, 205
network exploitation, 151
network mapping, 60, 131–43
networked devices, 251–53
neutrality, 211, 225–27, 239–40
Nieborg, David B., 251
No Shortcuts (Smeets), 270
nodes, 136
nongovernmental organizations (NGOs), 208, 217, 225–27
nonlinear thinking, 29–31, 35, 75, 204
nonproliferation, 183
nonstate actors, 201–2, 215–28
North Atlantic Treaty Organization (NATO), 197, 206–8
North Korea, 199–200
nuclear proliferation, 198

observation, 119–20
observe, orient, decide, act (OODA) loops, 123–24, 151
OECD (Organisation for Economic Co-operation and Development), 28
offensive tactics: choosing, 57; disruption, 148–58; feedback loops, 123–24; inflicting destruction, 180–85; mapping, 141–43; risk assessment, 79; types of, 60–61; value propositions, 93, 168
Olhava, Phoebe, 273–74